Yearbook of the Ayrshire Cattle Breeders For 1909
Including General Information on Ayrshire Cattle

by Ayrshire Breeders' Association

with an introduction by Jackson Chambers

This work contains material that was originally published in 1909.

This publication is within the Public Domain.

This edition is reprinted for educational purposes
and in accordance with all applicable Federal Laws.

Introduction Copyright 2017 by Jackson Chambers

Self Reliance Books

Get more historic titles on animal and stock breeding, gardening and old fashioned skills by visiting us at:

http://selfreliancebooks.blogspot.com/

Introduction

I am pleased to present another title in the "Cattle" series.

The work is in the Public Domain and is re-printed here in accordance with Federal Laws.

As with all reprinted books of this age that are intended to perfectly reproduce the original edition, considerable pains and effort had to be undertaken to correct fading and sometimes outright damage to existing proofs of this title. At times, this task is quite monumental, requiring an almost total "rebuilding" of some pages from digital proofs of multiple copies. Despite this, imperfections still sometimes exist in the final proof and may detract from the visual appearance of the text.

I hope you enjoy reading this book as much as I enjoyed making it available to readers again.

Jackson Chambers

Report of the Proceedings

——of the——

Thirty-fourth Annual Meeting

——of the——

Ayrshire Breeders' Association

——held at——

Hotel Manhattan, New York,

February 3, 1909.

The Thirty-fourth annual meeting of the Ayrshire Breeders' Association was held at Hotel Manhattan, New York, in response to a call from the Secretary, and was called to order at two p. m., by the President, E. J. Fletcher who said:

GENTLEMEN AND FELLOW MEMBERS:—The little I have to say to you today smacks of the same old sermons thought, care, and activity.

The past year has been a very trying one for all breeders of thoroughbred cattle on account of the financial condition of the country—But times are looking brighter, inquiries are coming in quite fast for better stock.

The dairymen, having to manufacture their products at a very small margin, are beginning to see their greatest profit is to breed *better dairy cattle,* breed a herd that will produce one third more at the same cost. They ask themselves the question: "How can this be done?" And they, at once, turn to the various dairy breeds to see which is best suited to their needs. They come to us to see what the Ayrshire can do. We must be ready to show them that the Ayrshire is best adapted to their requirements. And, when they come to us as *buyers,* we should be able to meet their questions with a convincing array of facts, and that means records.

The first thing they wish to know is: What kind of a record a cow can make under ordinary conditions. We should show them that 8000 lbs. of milk and 350 lbs. of butter is a better record, under the usual conditions than a much larger record under forced conditions, but, of course, we are pleased to experiment in order to learn the possibilities of the breed.

Last year over 15,000 lbs. of milk and over 750 lbs. of butter were produced from an Ayrshire cow—over 14,000 lbs. of milk and over 650 lbs. of butter from a four-year-old.

I was delighted with the records, nearly as much so perhaps, as the owners themselves. And I think we have many on our books capable of equaling those records, if the effort were made.

Another thing the dairyman wants to know is, not only the record of the cow's dam but *her own record.* It is the record of the animal and the records back of her that make the prices.

It has been proven in the other dairy breeds and it must follow in the Ayrshire. Therefore, we should be especially careful in our record-keeping, and I think the Association cannot put money out to any better advantage than to encourage the Ayrshire breeders to make tests for milk and butter.

The buyer also inquires as to their show ring winnings but his main point is "what can they do at the pail and what have their ancestors done?"

One thing he has made up his mind about: the cow must be nearly white. Well, the Chinese tell us they could tolerate the Americans, if they were not such a homely race. We would look better, they say, with yellow skins and eyes aslant. That is a matter of taste—as is the demand for nearly white cattle.

For myself, I prefer a finely marked red and white animal. But if the purchaser desires white let him have it. Color is a minor matter.

We should keep the Ayrshire before the public eye. We can do this by good cuts and descriptions in our best Agricultural papers, by public meetings, and by seeking the support of the Agricultural Colleges in our various states. We should endeavor to gain as strong a support from our government as the Canadian Government gives to our brothers across the border.

How this is to be done, is worthy of our thought and consideration. Our herds should show improvement from year to year. This can be done by making use of our previous experiences. A great many men have said: "You have the best breed of dairy cattle but they run short teated." I am sure this can be overcome by careful and proper breeding.

It is not best to push our breed by throwing mud at the other breeds—but let the animals speak for themselves.

The Ayrshire is fast coming to the front. There is a great demand for Ayrshire Stock, and in time they are bound to lead the dairy world.

We can truthfully say that our past year has been a prosperous one. Many sales have been made at satisfactory prices.

Our membership has been largely increased by the addition of 38 new members, and this means a great increase of help in our work throughout the country—and also shows the growth in popularity of the Ayrshire Breed.

Gentlemen. That our prosperity may continue let every member individually and unitedly, *think, speak,* and *act* for the interests of the Ayrshire Breeders' Association.

ROLL CALL.

Members Present.

Dorrance, Henry,	Plainfield, Conn.
Wells, Dudley, 2nd	Wethersfield, Conn.
Wells, William T.	Newington, Conn.
Gilbert, M. B.,	Wilmington, Mass.
Newton, L. W.,	Ashburnham, Mass.
Stickney, George E.,	Newburyport, Mass.
Bell, Charles J.,	Hollis, N. H.
Doud, Arthur V.,	Bristol, N. H.
Fletcher, Etna J.,	South Lyndeboro, N. H.
Hayes, Charles H.,	Portsmouth, N. H.
Strafford County Farm,	Dover, N. H.
The Uplands	Bridgewater, N. H.
Yeaton, George H.,	Dover, N. H.
Glen Alpine Farm	Morristown, N. J.
Lindsay, William	Plainfield, N. J.
Magie, J. D. & B. P.,	Elizabeth, N. J.
Probasco, W. V.,	Cream Ridge, N. J.
Ryan, P.,	Brewster, N. Y.
Karr, S. S. & Sons.,	Almond, N. Y.
Lansing, E. Ten Eyck,	Little Falls, N. Y.

Schanck, W. P., ...Avon, N. Y.
Sears, B. C. ....Blooming Grove, N. Y.
Tucker, W. G., ...Elm Valley, N. Y.
Zabriskie, Andrew C., ...Barrytown, N. Y.
Arkcoll, W. W. Blake, ...Paoli, Pa.
Butterfield, Jerome F., ... So. Montrose, Pa.
Deubler, James & Sons., ...Springvale, Pa.
Hillview Stock Farm, Limited, ...Paoli, Pa.
Oakey, John W., ...Bryn Mawr, Pa.
Winsor, Nicholas S., ...Greenville, R. I.
Copeland Farm ....Middletown, Vt.
Fletcher, A. M., ...Proctorsville, Vt.
Spalding, F. W., ...Poultney, Vt.
Stevens, C. B., ...St. Johnsbury, Vt.
Winslow, C. M., ...Brandon, Vt.

Represented by Proxy.

Macdonald, W. A., ...Mesa, Arizona
Aiken, Ella R., ...Norwalk, Conn.
Baton, John A. & Son.......... ...Wauregan, Conn.
Ennis, Alfred A....................Danielson, Conn.
Larned, J. H., ...Putnam, Conn.
Manwaring, John....................Norwich, Conn.
Palmer, Edward G., ...Plainfield, Conn.
Sears, N. E., ...Elmwood, Conn.
Wells, Dudley & Son..............Wethersfield, Conn.
Crabb, Frank A., ...Litchfield, Ill.
Bearce, George B., ...Lewiston, Maine
Buckley, J. P., ...Stroudwater, Me.
Blanchard, S. D. & Son...............Sanford, Me.
Dow, Fred N., ...Portland, Me.
Dunn, W. H. & G. H., ...Norway, Me.
Hunnewell, A. A., ...New Gloucester, Me.
McCrum, Lemuel, ...Mars Hill, Me.
Pember, Elmer F., ...Bangor, Me.

Scott, J. McPherson, ...Hagerstown, Md.
Bacon, P. K.,......................Campello, Mass.
Barnes, B. F.,....................Haverhill, Mass.
Burt, Jairus F.,.................Easthampton, Mass.
Copeland, Davis & Son..............Campello, Mass.
Doe, Charles C.,...................Lexington, Mass.
Easterbrook Brothers, ...Webster, Mass.
Essex County Training School, W. Grant
    Fancher, Supt.,................Lawrence, Mass.
French, C. A.,..................North Andover, Mass.
Leach, Philo, ...Bridgewater, Mass.
Marsh, William H.,:..............Barre Plains, Mass.
Peirce, F. C.,.................Concord Junction, Mass.
Pierce, George H.,...................Concord, Mass.
Piper, Anson C.,..................South Acton, Mass.
Sage, Charles D.,.............North Brookfield, Mass.
Smith, Peter D.,.....................Andover, Mass.
Stone, George F.,...................Littleton, Mass.
Tyler, Arthur F.,......................Athol, Mass.
Michigan School for the Deaf,............Flint, Mich.
Surget, James.......................Natchez, Miss.
University of Missouri................Columbia, Mo.
Abbott, J. N.,......................Concord, N. H.
Breck, Stephen R.,..................Claremont, N. H.
Childs, Harlow, N.,.................Piermont, N. H.
Clark, George C.,.....................Orford, N. H.
Cross, W. L., ...Ponemah, N. H.
Garvin, W. R.,........................Dover, N. H.
Hayes, Charles S.,.................Portsmouth, N. H.
Holt, Andy......................Lyndeboro, N. H.
Kimball, Herbert M.,..................Concord, N. H.
Rockwood, C. E. & Son...............Temple, N. H.
Russell, Frank E.,..................Greenfield, N. H.
Sawyer, E. E.,......................Atkinson, N. H.
Upham, Charles H. & Son.....Thornton's Ferry, N. H.

Beach, Frederick H., ... Morristown, N. J.
Tilton, E. A., ... Hamilton, N. J.
Whittingham, W. R., ... Milburn, N. J.
Arden Farms Dairy Co., Wm. Viner, Supt., Arden, N. Y.
Ballou, George William .... Middletown, N. Y.
Bell, George H., ... Rome, N. Y.
Bensley, M. F., ... Buffalo, N. Y.
Bentley, Ellis W., ... Windham, N. Y.
Bilby, Emerson .... Deposit, N. Y.
Burdick, George W., ... Friendship, N. Y.
Button, E. L., ... Melrose, N. Y.
Clark, C. W., ... Guymard, N. Y.
Clark, N. E., ... Potsdam, N. Y.
Converse, J. F., ... Woodville, N. Y.
Crowley, Thomas J., ... Potsdam, N. Y.
Griffin, J. H., ... Moira, N. Y.
Hall, Lott .... Gouverneur, N. Y.
Hamilton, William Pierson .... Sterlington, N. Y.
Hatch, C. E., ... Gainesville, N. Y.
Hill, J. Edwin & Son .... Gouverneur, N. Y.
Leach, J. S. & Son .... Gouverneur, N. Y.
Lewis, C. W. & Sons .... Alfred Station, N. Y.
Litchard, A. L. & Son .... Rushford, N. Y.
McCrea, Robert .... Champlain, N. Y.
Ormiston, James .... Cuba, N. Y.
Pike, George E., ... Gouverneur, N. Y.
Ricker, Clarence .... Belmont, N. Y.
Rogers, G. L., ... Gouverneur, N. Y.
Siver, D. E., ... Cooperstown, N. Y.
Smith, Oliver & Son .... Chateaugay, N. Y.
Stetson, Francis Lynde .... Sterlington, N. Y.
Stowell, F. D. & E., ... Black Creek, N. Y.
Stowell, W. C., ... Black Creek, N. Y.
Taber, George .... East Aurora, N. Y.
Tubbs, Ambie S., ... Maple View, N. Y.

Underhill, C. S., ...Glenham, N. Y.
Welch, M. G. & Son...................Burke, N. Y.
Whipple, L. W. & Son................Malone, N. Y.
Will, John........................ Ft. Covington, N. Y.
Davison, Milton W., ...Canisteo, N. Y.
Betts, Henry..........................Pittsfield, Ohio
Cook, Howard..........................Beloit, Ohio
Greenawalt, J. S. & Son.................Beloit, Ohio
Blakeslee, O. P., ...Spartansburg, Pa.
Byrne, Christopher................ .Friendsville, Pa.
Cass, George L., ...New Milford, Pa.
Davis, Edward P., ...Newtown, Pa.
Friends Asylum, Frankford.........Philadelphia, Pa.
McFadden, George H., ...Bryn Mawr, Pa.
Peck, C. L., ...Nelson, Pa.
Roberts, Percival, Jr., ...Narberth, Pa.
Templeton, Robert & Son..................Ulster, Pa.
Turnbull, Thomas Jr., ...Allegheny, Pa.
Valentine, John R., ...Bryn Mawr, Pa.
Bowen, Edward S., ...Pawtucket, R. I.
Brown, Obadiah, Estate of..........Providence, R. I.
Sherman, Everett B., ...Harrisville, R. I.
Smith, Daniel A., ...Tarkiln, R. I.
Tefft, S. Frank.......................Hamilton, R. I.
Vaughn, William P., ...Providence, R. I.
Hinson, W. G., ...Charleston, S. C.
Cosgrove, Michael...............Madison, So. Dakota
Groome, H. C., ...Warrenton, Va.
Venable, A. R. Jr., ...Farmville, Va.
Abell, C. A., ...St. Albans, Vt.
Anderson, A. J. & Son..........North Craftsbury, Vt.
Buck, C. W., ...Brownsville, Vt.
Butterfield, B. F., ...Derby Line, Vt.
Clark, H. A., ...Hyde Park, Vt.
Collins, F. O., ...St. Albans, Vt.

Cramton, W. S., ...Rutland, Vt.
Davidson, George....................So. Royalton, Vt.
Drew, F. A., ...So. Burlington, Vt.
Dunsmore, George......................Swanton, Vt.
Emerson, Charles W., ...Charlotte, Vt.
Forest Park Farm, ...Brandon, Vt.
Foss, J. Barron......................St. Albans, Vt.
Hannah, Matthew.....................Brownsville, Vt.
Houghton, W. W., ...Lyndonville, Vt.
Joslyn, F. A., ...Northfield, Vt.
Lovejoy & Eddy...........................Stowe, Vt.
Merriam, W. A., ...Glover, Vt.
Parker, R. & Son.....................Ferrisburg, Vt.
Proctor, Fletcher, D., ...Proctor, Vt.
Rice, George L., ...Rutland, Vt.
Scott, W. F., ...Brandon, Vt.
Scribner, G. S., Estate of..............Castleton, Vt.
Turner, Walter D., ...Moretown, Vt.
Vaughan, C. A. & R. H., ...Thetford, Vt.
Vermont Industrial School.............Vergennes, Vt.
Watson, H. R. C., ...Brandon, Vt.
LaDue, William......................Braintree, Vt.
Martin, H. W., ...Bradford, Vt.
Soule, Arthur B., ...St. Albans, Vt.
Clise, J. W., ...Seattle, Washington
Reymann, L. A., ...Wheeling, West Va.
McMillan, Gilbert..............Redmond, Washington
Finn, James........................Whitewater, Wis.
Nelson, George A., ...Milltown, Wis.
Seitz, Adam........................Waukesha, Wis.
Tschudy, Fred ....Monroe, Wis.
Hunter, Robert & Sons................Maxville, Ont.
Ness, R. R., ...Howick, Que.
Stephen, W. F., ...Huntington, Que.

REPORT OF SECRETARY

The Ayrshire Breeders' Association has, during the past year, made steady progress in bringing to the front the dairy ability of the Ayrshire cow.

We have during the past year brought out a champion quintette that is ahead of any official record of any Ayrshire cows or heifers in the world.

We have reason to congratulate ourselves that the result of our dairy tests have placed the Ayrshire cow of America as the champion Ayrshire cow of the world, and that Rena Ross by her wonderful record gives character to the Ayrshire breed as a dairy cow.

FIVE CHAMPION COWS

Baby Douglas owned and tested by L. A. Reymann, Wheeling, W. Virginia, gave a two year old record of 9,652 lbs. of milk and 440 lbs. of butter.

Madonna Lass 3d owned and tested by L. A. Reymann, gave a three year old record of 10,467 lbs. of milk and 502 lbs. of butter.

Jennie of Sand Hill owned and tested by S. S. Karr & Sons, Almond, N. Y., gave a three year old record of 10,160 lbs. of milk and 510 lbs. of butter.

Bessie of Rosemont owned and tested by Geo. H. McFadden, Bryn Mawr, Pa., gave a four year old record of 14,102 lbs. of milk and 675 lbs. of butter.

Rena Ross owned and tested by John R. Valentine, Bryn Mawr, Pa., gave a mature cow's record of 15,072 lbs. of milk and 751 lbs. of butter.

These records conducted by the Experiment Stations, and for a full year, are of great value as showing, not only the producing ability of the Ayrshire, but her staying quality.

A seven day, or a thirty day test is interesting, but a 365 day test shows what a cow can do.

ADVERTISING THE BREED

We issued as attractive a year book as possible, and then advertised it free on application, inserting a short advertisement in a number of papers, which we believe did a large amount of advertising for a little money for from the day the notices appeared in the papers, to this day, we have been constantly asked for copies of the year book.

In addition to this we have constantly received letters asking about the Ayrshire cattle, which we have always answered as carefully as possible, and also sent printed information.

Since the last meeting we have issued Volume XIX Ayrshire Record, 518 pages, and have also closed Volume XX which is now in the printer's hands, and will probably have about as many pages as Volume XIX and nearly 500 already recorded in Volume XXI.

We are on the last 25 in reaching 25,000 cows, and some 11,600 bulls.

SHOWING AT THE FAIRS

The showing of Ayrshires at the fairs was a credit to the breed, and at two fairs, in Maine and Oregon, the dairy tests showed the Ayrshire in the front rank as a dairy cow. I think the cows winning were owned by Mr. Ness of Auburn, Maine, and Mr. Clise of Seattle, Washington.

DEATHS

We have to record the death of Franklin P. Clark, Sudbury, Mass., one of the firm of Ormiston Brothers,

Cuba, N. Y., N. H. Winter, Cortland, N. Y., A. S. Shimer, Redington, Pa., and Charles Sanford, Ludlow, Vermont.

AYRSHIRE AUCTION

The dispersion sale of the Forest Park Herd of Ayrshires at Brandon, Vt., occurred as advertised.

It was under the management of the noted Leander F. Herreck, which in itself was a guarantee that everything had been done to make the sale a satisfactory one to both seller and buyer.

The cattle were in fine condition, with a good many choice animals. It was a dispersion sale with no reservations and no "by-bidders," honestly advertised and honestly sold.

There were 95 head sold, and what was a good recommend for the breed in general was that out of so large a number there was only one that had any defect in her udder. There was one three teated heifer.

Ten head sold for $2340.00, 25 head sold for $4525.00 and 37 head brought from $100.00 upwards.

Fifteen of the animals sold were young calves dropped since September, which averaged $33.00 each, bulls and heifers.

NEW MEMBERS

C. A. & R. H. Vaughan,	Thetford, Vt.
Adam Seitz,	Waukesha, Wis.
Charles H. Upham & Son,	Thornton Ferry, N. H.
F. O. Collins,	St. Albans, Vt.
S. S. Story,	North Stockholm, N. Y.

James Finn,	Whitewater, Wis.
J. Edwin Hill & Son,	Gouverneur, N. Y.
John Manwaring,	Norwich, Conn.
James G. Ray,	Unionville, Mass.
F. L. Haseltine & Sons,	West Union, N. Y.
Geo. E. Stickney,	Newburyport, Mass.
S. D. Blanchard & Son,	Sanford, Maine
J. B. Silliman,	Kortright, N. Y.
L. W. Newton,	Ashburnham, Mass.
J. D. Farrell,	Seattle, Wash.
Gilbert McMillan,	Redmond, Wash.
Wayside Farm,	Putnam, Conn.
R. J. Averill,	Washington Depot, Conn.
O. W. Brusie,	Millerton, N. Y.
James S. True,	Lewiston, Maine
W. A. McDonald,	Mesa, Arizona
H. W. Martin,	Bradford, Vt.
H. C. Copeland,	Middletown, Vt.
Geo. A. Nelson,	Milltown, Wis.
Fred N. Dow,	Portland, Maine

FINANCIAL REPORT OF SECRETARY

By Entries to Members	$1467	10
" Entries to Non-members	568	25
" Transfers to members	102	80
" Transfers to non-members	61	00
" Customs certificates	4	00
" Duplicate certificates	2	50
" Private Herd Books	10	50
" Milk record blanks	10	48
" Pedigree blanks	7	50
" Advertising in Year Book	78	00
" Volume XVIII	2	25

By Volume XVIII	$ 2	25
" New members	625	00
" Received from contestants in tests	116	89
	$3,058	52
" Received from N. S. Winsor, Treas. to balance account	848	97
	$3,907	49

Approved January 19, 1909, Geo. H. Yeaton, Auditor.

To Postage	159	65
" Express and freight	20	54
" Telegraph and telephone	9	27
" Home Dairy Test prizes 1906-7	210	00
" Home Dairy Test prizes 1907-8	210	00
" Banquet at Fifth Avenue Hotel	38	00
" Stenographer at annual meeting	57	25
" Surety bond for Treasurer	12	00
" Corporation tax	10	00
" Safe	75	00
" Exchange on typewriters	67	25
" Advertising in Country Gentleman	100	00
" Advertising in Hoard's Dairyman	100	00
" Advertising in Breeder's Gazette	9	80
" Advertising in Kimball's Dairy Farmer	8	40
" Advertising in Practical Dairyman	9	08
" Advertising in Ruralist	2	00
" Empire Engraving Company	105	29
" Derby Silver Company, engraving cup	3	00
" Engrossing certificates	4	00
" Argus Co., for transfer book	7	00
" Tuttle Co., for year book	304	00

To Sundries, ink, typewriter ribbons, rubber
       bands $2 45
" Travel, personal 114 03
" Travel and expense of Committee....... 96 36
" Trunk 7 00
" Argus Co., for certificates............. 4 50
" Argus Co., for book of certificates...... 10 00
" Brandon Publishing Co., printing...... 130 00
" Woodruff, printing.................... 1 90
" Experimenting Stations for testing...... 1019 72
" Salary of Secretary................... 1000 00

$3,907 31
18

$3,907 49

Approved Jan. 19, 1909, Geo. H. Yeaton, Auditor.

TREASURER'S REPORT.

Balance on hand, Jan. 1, 1908............. $3,531 81
    Receipts:
May dividend, Smithfield Sav. Bank......... 48 64
Nov. dividend, Smithfield Sav. Bank........ 49 60
July dividend, Rutland Sav. Bank.......... 27 46
Jan. dividend, Rutland Sav. Bank.......... 27 11
Interest on deposits, (C. M. W. ck.)........ 50 00
Sale of books............................ 122 25

$3,856 87

Payments per vouchers:
Expense, Treas. to Annual Meeting $19 50
Expense, Auditor................   28 90

Postage stamps	$6 50	
Freight on books	9 68	
From French fund	38 60	
From French fund	47 10	
Paid Secretary	848 97	999 65

$2,857 22

Balance due on the French fund.......... 1,576 24

Total Cash and bank deposits............. $4,433 46

Dover, N. H., Feb. 1, 1909.

This certifies that I have examined the account of the Treasurer of the Ayrshire Breeders' Association for the year 1908 and find credit given for all money received, vouchers for all payments and the account correctly balanced.

GEORGE H. YEATON,
Auditor.

INVENTORY IN OFFICE OF SECRETARY
DECEMBER 31, 1908.

1 index card and letter case	$50 00
1 Writing desk and typewriter combined	45 00
2 Typewriters	150 00
1 Letter copy machine and desk	35 00
1 Burroughs adding machine	250 00
147 Private Herd Books	147 00
Postage stamps on hand	1 65
31 volumes Scotch Herd Books	31 00
16 volumes Canada Herd Books	16 00
4 volumes Bagg Herd Books	4 00
4 volumes Sturtevant Herd Books	4 00
Usual supply of stationery and blanks	....
1 Safe	50 00
	$783 65

INVENTORY OF BOOKS IN HANDS OF TREASURER, January 1, 1909.

	Need Rebinding.	Good Condition.
Vol. 1		150
Vol. 2 (Old Edition)		2
Vol. 2 (New Edition)		94
Vol. 3		84
Vol. 4	3	114
Vol. 5	5	14
Vol. 6		184
Vol. 7		185
Vol. 8		193
Vol. 9		212
Vol. 10		209
Vol. 11		207
Vol. 12		218
Vol. 13		225
Vol. 14		221
Vol. 15		237
Vol. 16		245
Vol. 17		188
Vol. 18		257
Vol. 19		273
	8	3,512

3,512 Volumes at $2 00 each.............$7,024 00
8 Volumes in bad condition at $1 00... 8 00

Total value of volumes.................$7,032 00

GEORGE H. YEATON,
Auditor.

REPORT OF THE
HOME DAIRY TEST COMMITTEE FOR 1907-8

The year's test is from April 1, 1907 to March 31, 1908 inclusive.

SINGLE COW PRIZES.

First Prize of $30.00 to Rena Ross 14593, owned by John R. Valentine, Bryn Mawr, Pa. Calved June 15, 1907. Record, 12,937 lbs. of milk and 653 lbs. of butter.

Second Prize of $20.00 to Brown Eyes of Knockdon 19,216, owned by George H. McFadden, Bryn Mawr, Pa. Calved April 26, 1907. Record 11,328 lbs. of milk and 505 lbs. of butter.

Third Prize of $10.00 to Queen of Barclay 15,096, owned by George H. McFadden, Bryn Mawr, Pa. Calved March 24, 1907. Record 11,158 lbs. of milk and 480 lbs. of butter.

HERDS OF FIVE COWS FOR BUTTER

First prize of $75.00 to the herd owned by J. F. Butterfield, South Montrose, Pa.

	MILK	BUTTER
Bernice Sebastian 20528		
Calved May 16, 1907..........	8,935	389.73
Abbie Sebastian 20531		
Calved May 16, 1907..........	9,497	384.37
Hazel Sebastian 20530		
Calved May 12, 1907..........	7,755	380.59
Ruth Webb 17457		
Calved May 10, 1907..........	7,982	370.46
Pauline Sebastian 18678		
Calved July 9, '06-Oct. 16, '07..	8,871	363.27
	43,040	1,888.42

Second prize of $50.00 to the herd owned by George H. Yeaton, Dover, N. H.

	MILK	BUTTER
Bert Gyna 16570		
Calved Mar. 1, '07-Jan. 14, '08..	9,670	399.13
Maumee 16566		
Calved Oct. 7, '06-Oct. 24, '07..	8,846	370.89
Ouilma 16564		
Calved Dec. 1, '06-Oct. 13, '07..	8,268	364.76
America 16572		
Calved June 8, '07............	6,921	315.33
Gebic 13981		
Calved July 22, '06-June 15, '08	6,688	311.82
	40,393	1,761.93

Third prize of $25.00 to the herd owned by George F. Stone, Littleton, Mass.

	MILK	BUTTER
Vera 16154		
Calved Sept. 4, '06-Sept. 1, '07	8,765	381.79
Crinkle Corslet 16683		
Calved before April 1907......	8,975	345.78
Lily Carlton 21110		
Calved May 19, '07...........	6,107	299.06
Crimson Rambler 21109		
Calved Aug. 22, '06-Dec. 12, '07	6,985	295.91
Esther Carlton 21786		
Calved April 25, 1907........	6,609	277.54
	37,441	1,600.08

The French prize cup valued at $75.00

This prize is for both milk and butter, as determined by points. Won by the herd owned by George H. McFadden Bryn Mawr, Pa.

	MILK	BUTTER
Bessie of Rosemont 17904		
Calved August 1, 1907.........	10,377	477.79
Frisky of Bonshaw 17018		
Calved Mar. 24, '07-May 2, '08	9,353	450.63
Lizzie of Barclay 17024		
Calved Mar. 28, '07-Feb. 26, '08	8,858	429.86
Clockston Bella 2d 21628		
Calved March 1, 1907.........	8,509	420.86
Broomhill Minnie 10th 21627		
Calved May 21, '06-Aug. 24, '07	7,921	406.06
	45,018	2,185.20

Points for milk and butter, allowing one point for each pound of milk and 17.5 points for each pound of butter, 48,842 points.

C. M. WINSLOW   W. V. PROBASCO

THOMAS TURNBULL, JR.

Committee on Home Dairy Tests.

THE PRESIDENT—Unless there is some objection to the Home Dairy Test Committee's report as read, it will stand approved.

THE SECRETARY—The Executive Committee recommend that the time of commencing the year's test in the Home Dairy Test be changed from April 1st to October 1st and in order to avoid an interim of six months in which no test is being carried on, that for the year 1909 there shall a test begin April 1st, and another begin October 1st, each of them to run one year from time of beginning.

A member who enters his herd for beginning at the April, may also enter for the October, either with the same herd wholly or in part or with an entirely different herd.

After the 1909-10 tests are finished, there shall be only the test carried on that begins in October of each year.

Mr. Oakey—I move the adoption of the resolution as read.

Resolution adopted unanimously.

The Secretary—The Executive Committee recommend a change in the prizes offered in the Home Dairy Test, the change to take effect at the beginning of the next year's tests for April and October, and to read as follows:

For Herds of five cows each, six prizes:—1st prize a silver cup valued at $75, 2nd prize $50, 3rd prize $40, 4th prize $30, 5th prize $20, 6th prize $10.

For individual cows, six prizes:—1st prize $30, 2nd prize $20, 3rd prize $15, 4th prize $10, 5th prize Diploma Highly Commended, 6th prize Diploma Commended.

The President—You have heard the resolution, Gentlemen, what will you do with it. Has anyone any remarks to make. If not, I await a motion.

Mr. Oakey—I move the adoption of the resolution as read.

Resolution unanimously adopted.

The Home Dairy Test Committee of last year were re-elected, to serve for the coming year, viz: C. M. Winslow, Dr. Thomas Turnbull, Jr., and W. V. Probasco.

EXPERT JUDGES.

The following list of expert judges were selected by the Association at the last annual meeting, and recommended as competent to judge Ayrshires in the ring at all fair associations.

E. J. Fletcher....................Greenfield, N. H.
George H. Converse................Woodville, N. Y.
W. F. Stephen....................Huntington, Que.
Prof. M. A. Scovill..................Lexington, Ky.
Gilbert McMillan...................Redmond, Wash.
Prof. W. L. Carlyle...................Denver, Col.
G. Arthur Bell....................Washington, D. C.
George H. Yeaton......................Dover, N. H.
L. A. Reymann...................Wheeling, W. Va.
Prof. Harry Hayward..................Newark, Del.
Prof. C. S. Plumb..................Columbus, Ohio
Howard Cook...........................Beloit, Ohio
George E. Stickney ....Newburyport, Mass.

MR. STICKNEY—I am no judge of Ayrshire cattle, I assure you.

MR. HAYES—No, he's no judge. He bought some of me.

MR. OAKEY—I heard this morning that he was a good judge of Ayrshire cattle, and I recommend that his name remain on the list.

DR. BUTTERFIELD—I move the list as read be adopted and the Secretary be instructed to send the list to all the Fair Associations where Ayrshire cattle are likely to be exhibited.

Motion seconded and unanimously adopted.

THE SECRETARY—The Executive Committee recommend that after this test now being conducted is ended, the awards be made on a basis of both milk and butter, and be decided by a scale of points, as follows: One point for each pound of milk credited to a cow, and 17.5 points for each pound of butter fat reported by the Experiment Station, obtained from each cow.

THE PRESIDENT—What will you do with this recommendation of the Executive Committee.

MR. OAKEY—I move the adoption of the recommendation as read.

Voted unanimously.

THE SECRETARY—The Executive Committee recommend a series of champion prize ribbons for the cow that shall give the highest record for milk and butter fat, in a continuous test from two to five consecutive years, the first ribbon being given for a two year championship, then if she is able to hold the leadership continuously for three, four and five years, she shall receive a championship ribbon for each year.

This shall be conducted through the Home Dairy Test, beginning October 1st, 1909, and a new champion may start each year, thereafter.

MR. BELL—I would second the resolution, as it will bring out the long milking cows, and will be of no material cost, and will not conflict with anything.

MR. OAKEY—I approve the resolution, and would also second it.

THE PRESIDENT—Those in favor of the resolution as read please say aye. It is a unanimous vote.

THE SECRETARY—It is recommended by the Executive Committee that the Association appropriate the sum of $300.00 to each of the two Expositions, the Alaska Yukon-Pacific Exposition and the National Dairy Show both of which are to be held this fall.

The President—You have heard the recommendation. Has anyone anything to say upon the question?

Mr. Dodge—I move the adoption of the resolution, but I think it should be for $500.00 each instead of $300.00.

The National Dairy Show is considered the greatest show of dairy cattle in America, and I think the Ayrshire Breeders' Association can afford to appropriate from $500.00 to $1,000.00 as a premium to bring out a large representation of the breed.

There are two ways of advertising a breed.

One is to show them at fairs, and the other is to show them at home, but a great many people go to fairs who never take the trouble to read papers, and we want to hit both classes.

The Alaska Yukon-Pacific Exposition is going to be a large affair, and is held in a country where Ayrshires are just beginning to be appreciated, and I believe this Association cannot spend money to any better advantage than to send it out there for premiums.

I believe $1,000.00 would be well spent there, and I move that we appropriate $1,000.00 to each of the two fairs, the Alaska Yukon-Pacific Exposition and the National Dairy Show.

Mr. Bell—I do not doubt but that $1,000.00 at either place would be well spent, but where is the $2,000.00 coming from.

As I understand our financial condition we are not in the shape to give as much as that, and I believe $300.00 each is all we can afford to appropriate.

If some way can be devised whereby we can raise more than that without taking it from our Treasury I should heartily approve, for I endorse all that has been said about the advantage of making a large show of Ayrshires at both places.

Mr. Spalding—I agree with a large part of what Mr. Dodge has said, but our finances will not allow us to compete with the treasury of the Jersey Cattle Club, or the Holstein Association.

If we appropriate what we are able, we show our appreciation of their part of the country, and what they are doing, and show our good will.

Mr. Hayes—Perhaps the gentleman can suggest some way of getting the needed money. It is easy to spend money, but hard to accumulate it. We want to keep within our means and be respectable.

The President—Those in favor of appropriating $300.00 to be applied as special premiums in the single animal classes at each of the two fairs, the Alaska-Yukon-Pacific Exposition, and the National Dairy Show make it manifest by saying Aye, and contrary by saying No.

The Ayes have it and it is a vote.

Mr. Sears—I would move that a committee of three be appointed at this meeting to try and raise $500.00 to be used as a special prize for herds shown at the Alaska-Yukon-Pacific Exposition, and that the money raised be divided pro rata between all the herds shown in the ring.

Mr. Converse—Do I understand that this money may be competed for by Ayrshires from other countries.

The Secretary—In order to be eligible to compete for the Association special prizes, the animals should be registered in the Ayrshire Register, and the exhibitor be a member of the Ayrshire Breeders' Association.

The President—Those in favor of the appointment of a committee of three by the chair to raise a fund for special prizes at the Alaska-Yukon-Pacific Exposition please say Aye, those opposed say No. It is a unanimous vote, and I will appoint Mr. Sears, Mr. Oakey and Mr. Gilbert.

Mr. Yeaton—We have stored in Greenville, R. I., at the home of the Treasurer some over 3500 volumes of the Herd Books. Last year it was voted to have them put in cases and insured. I would like to ask Mr. Winsor to state to the meeting the condition of the matter.

Mr. Winsor—Upon investigation I found it would cost over $400.00 for storage, and that it would not lessen the cost of insurance to a corresponding degree, so I waited for further instructions.

Mr. ....—I would like to make a motion that three full sets be placed in some safe deposit, and the balance sent to the Secretary; also that those that are placed in deposit be at the disposal of the Secretary of the Association.

Mr. Yeaton—There is no danger of the herd book being destroyed so effectually as to prevent our obtaining a set for renewal even if all the new books should be destroyed, for there are a great many full sets scattered all over the country, in the hands of the members of the Association.

Mr. Converse—Why is it necessary to send these two or three thousand volumes to the Secretary when they are insured where they now are? It seems useless to send them to Vermont.

Mr. Gilbert—If a person wishes to inquire about the herd book, he naturally writes to the Secretary.

All business should be done under the direction of the Secretary, and I would amend the motion by having all the herd books shipped to the Secretary and placed in his care.

The President—Those in favor of the amendment say Aye, those opposed say No.

It is a vote, and the amendment is carried.

Will you instruct the Treasurer to ship the books to the Secretary, to be under his care?

Those in favor say Aye, those opposed no, and it is a vote.

THE PRESIDENT—The next business in order is the election of officers. I will appoint George E. Stickney, William T. Wells and George H. Yeaton as inspectors of proxies, and tellers. They reported that there were 145 proxies divided among twenty-two members, with holdings ranging from one to nineteen, and that each member had a right to his own vote and as many additional votes as he held proxies for.

MR. WM. T. WELLS—In the election of officers I move that the Secretary be empowered to cast one vote for Mr. E. J. Fletcher to become president for the ensuing year.

THE PRESIDENT—You have heard the motion of Mr. Wells. Those in favor say Aye, those opposed say no, and it is a unanimous vote.

The Secretary declared Mr. Fletcher elected unanimously.

THE PRESIDENT—I thank you gentlemen, and I assure you I will try to do my best to push the Ayrshires to the front.

MR. SCHANCK—I move the Secretary cast one ballot for J. F. Converse for Vice-President, which was done and Mr. Converse declared elected.

MR. ....—I move the Secretary cast one ballot for Mr. J. W. Clise for second Vice-President.

MR. STICKNEY—I object, not to the man, but to the method. There are four vice-presidents to be elected, and I think the Association as an Association should have the privilege of voting for the officers.

Mr. Bell—I would put in nomination Mr. George E. Pike as second vice-president.

Mr. Wm. T. Wells—Are there two nominees for this office?

The President—Yes. Mr. Clise and Mr. Pike.

Each gentleman who holds proxies can vote his own vote and whatever proxies he holds.

Mr. Wm. T. Wells—We all know Mr. Clise is doing splendid work for the breed, and I strongly favor Mr. Clise's election.

Mr. Stickney—I have no objection to Mr. Clise, and approve our having him as our representative in the west.

Mr. Winslow—I am sorry these two men are placed in opposition, for I approve of both, and would like to see both elected as vice-presidents.

Mr. ——Why should we vote proxies when we have a quorum.

Mr. Stickney—Why should we not vote proxies? Members who could not be present have sent their proxies expecting their representative would vote for them, and it is right they should. Every member of the Association has a right to vote either in person or by proxy.

Mr. ....—I am sorry this thing has come up, for Mr. Pike is at home on a sick bed. He has done as much for the Association as any other member, and we should not throw him down when he lies at home on a sick bed and unable to be present with us.

Mr. ....—I am a stranger in this meeting, but one of the few points in the Ayrshire Association is the lack of advertising the breed. I am now with a man who has a Guernsey herd. He is of the opinion that the Ayrshires are a good breed but not advertised enough. If you can assist your Association by naming Mr. Clise, by all means name him for your Association.

Mr. Hayes—We are not talking about advertising now; we are talking about vice-presidents.

Mr. Gilbert—I do not see anything that should hinder our considering both Mr. Pike and Mr. Clise.

Mr. Stickney—I don't think anyone has any objection to Mr. Clise. There is plenty of room for both Mr. Pike and Mr. Clise.

Mr. Spalding—The Association originally started in the East, but our family is enlarging, and what we want is to keep the East and West united and all pull together.

Mr. Gilbert—I make a motion that the Secretary be empowered to cast a ballot for Mr. Pike for Second Vice-President, and for Mr. Clise for Third Vice-President.

Mr. Lansing—I second the motion of Mr. Gilbert. I am from Central New York, and it strikes me that perhaps the Ayrshire breeders have been too much of a close corporation, and should stretch our lines westward.

The President—Those in favor of the motion say Aye; contrary, No. And the Secretary casts a vote for Mr. Pike for Second Vice-President and Mr. Clise for Third Vice-President, and I declare them elected.

Mr. Schanck—I move that the Secretary cast one ballot for Mr. J. A. Ness of Auburn, Maine, as Fourth Vice-President.

This was done and the President declared him elected.

Mr. Stickney—If there be no objection, I would be most heartily pleased to place in nomination for Secretary a man who has for a good many years held the office, Mr. C. M. Winslow.

Mr. Gilbert—I would like to second the nomination of Mr. Winslow. Unless there is good reason for a change, it is poor policy for any organization to change its Secretary, and especially an organization of this kind, where so much is required of a Secretary, and such varied work, for it is a fact that if the work is done fairly well, the longer he remains in the office, the more valuable he becomes from the knowledge he constantly acquires of the work. It would take a new man four or five years to learn the work, and here we have a man who is thoroughly acquainted with the work in all its detail, and he can do more for the prosperity of the Association than any new man can do. I not only recommend his re-election, but I also recommend an increase of salary.

Mr. Dodge—I appreciate the gentleman's remarks, but while I am not a member of your Association, I represent one of the large herds, and while I have a good opinion of your present Secretary as a man and a breeder, and admire the work he has done for the Association and the breed in the past, I believe our Association is growing, and to place the Ayrshire cow where she belongs it must go faster. We need new blood in this Association if we are going to make strides and place the Ayrshire cow where she belongs. I know a young man who will work for the breed. He will push the breed, and I place before you the name of Mr. Wm. T. Wells of New England.

Mr. Oakey—There is a great deal of talk about this question, and I rise to second the name of W. T. Wells. I have the pleasure of being a member of the Association, also of representing one of the best herds. There is no one who appreciates the Secretary more than I do, but I think if we could get a younger man to help the people to carry the wood and water, to help them along, to encourage them, it would mean a great deal to the Association. At the National Dairy Show last fall, we four men who were there felt lonesome. Every possible effort was made by the other dairy breeds to push their respective breed, and the Ayrshire was without a representative. The Holstein, the Jersey, and the Guernsey people had representatives spreading the gospel of their breed far and wide over that great assembly, and nothing doing for the Ayrshire. Another thing—there are a number of men this year who could not start their cows in the advanced registry by reason of not being able to hear from the Secretary in time to start. As we have started this thing, we might as well have the sense of the meeting, but whoever is elected Secretary will be my Secretary and I will stand by him.

Mr. Gilbert—By what means complaints have got into the papers I do not know, but I think it very bad taste for any one to have them printed in any newspaper, for the better way is to bring them right here. As far as my business with the office is concerned, everything has been done promptly and satisfactorily. It is not a question of age or change of Secretary. There may have been a press of business, and consequent neglect to respond quickly to inquiries, and that is the reason why I spoke of his needing more money to do with. What was it that enabled the other breeders to so satisfactorily represent their breeds at the National Dairy Show at Chicago? It was money, and you can never establish any breed on this continent until you get them into the hands of monied men. We are not going to do it by changing Secretaries. I have no objection to Mr. Wells. I do not know who he is, but I say let well enough alone. If you have any criticism about the Secretary, put them up and let us dig them over. Newspapers are just as liable to make mistakes as anybody else. It is a question of business. If you have anything to say, cough it up and let us right it.

Mr. Spalding—I have been connected with Ayrshires about as long as any man here with one exception. I am familiar with the workings from the start. I have been in the Secretary's office morning, noon and night. I have been acquainted with Mr. Winslow for a long time, and I believe everything has been done for the benefit of the Association. The Association is growing with a healthy, steady growth, and we want things done in the future in the same honorable, progressive way as they have been in the past. In view of the reliable work the Secretary has always given us, and in view of his character and standing, I move you, gentlemen, a second nomination of Mr. C. M. Winslow.

Mr. J. D. Wells—I would like, without casting any aspersions on Mr. Winslow and assuring you that we have been his staunch supporters, to second the nomination of Mr. Wells. I think the same trouble prevails in this Association that does in our herd of Ayrshires at home. My father's herd is one of the oldest herd of Ayrshires in the country. We get no new blood, and we are behind. It needs some new blood. This Association needs new blood to start it, and let us have new blood.

Mr. Bell—As I understand it, the complaint is that our Secretary has not done quite as much as the Secretaries of some other breeds, but I would like to ask how you are going to expect a man to accomplish as much on a salary of $1,000 as another can on $2,800. I cannot see where we are going to be benefited by putting in anyone else for the same money. If you would give this man what the Holstein breeders pay their Secretary, you might have some fine work from him, but we are not in a position to do this. I think we have had a good showing the last few years, and I think it will continue to improve.

MR. GILBERT—I do not wish to impose on you, or take too much time, but there is one thing to be considered in trying to raise the reputation of the breed. We have got to have men in office with established reputations. I think that is one of the strongest arguments in favor of Mr. Winslow . There is no man that can dispute but what Mr. Winslow is considered a square, upright man and a man that understands the Ayrshire breed. We have a man that the public has confidence in, and let us have confidence in him and keep him where he is.

THE PRESIDENT—Are you ready for the ballot?

MR. STICKNEY—The list prepared by the Committee of Three gives the number of votes each member is entitled to cast, which can be read and the person can announce for whom he casts his votes.

This was done and the tellers declared 136 votes for Mr. Winslow and 36 votes for Mr. Wells.

Mr. Wm. T. Wells moved that the election of Mr. Winslow be made unanimous, which was seconded by Mr. Dodge, and passed.

MR. WINSLOW—I thank you, gentlemen, and I want to say that I am the Secretary of every member of the Association, just the same as I always have been.

MR. HAYES—You want to circulate around a little more.

MR. LANSING—What is the salary of the Secretary?

THE SECRETARY—My salary began at $400, but at that time we had only $800 in the treasury. A few years ago it was increased to $700, and last year to $1,000, and I have always furnished the office free of charge.

Mr. Lansing—I move the salary be increased 20 per cent.

Mr. Hayes moved an amendment, seconded by Mr. Oakey, that the salary be $1,500. Carried, and the salary increased to $1,500.

The Secretary—I thank you, gentlemen, more for the Association than for myself personally, for it will relieve me from considerable routine desk work and give me time to do work in the line of pushing the breed.

Mr. Wells—I almost feel that the candidate has accomplished the work we had in view.

The President—Who will you have for Treasurer?

Mr. Spalding—I move the Secretary cast one ballot for Mr. N. S. Winsor.

Motion seconded and carried, and Mr. Winsor was declared elected Treasurer.

Mr. ....—I move that the Secretary cast one ballot for Mr. Geo. H. Yeaton for Auditor.

Motion seconded and carried, and Mr. Yeaton was declared elected Auditor.

Mr. Spalding—I move the Secretary cast one ballot for Mr. Schanck to succeed himself.

Motion seconded and carried, and Mr. Schanck was declared elected.

Mr. Wm. T. Wells—I move that the Secretary be instructed to cast one ballot for Mr. Howard Cook to succeed himself.

Motion seconded and carried, and Mr. Cook declared elected.

NAMES FOR MEMBERSHIP.

P. T. Fitzgerald and George R. Wales were elected members.

MR. STICKNEY—I would like to make a statement which I wish kept from publication in the papers. I would like to donate a silver cup to the Association, to be offered in the name of the Association, the particulars of which I leave with the Executive Committee to arrange.

(*Note by the Secretary.*—This cup which Mr. Stickney so modestly presents to the Association is now in my hands, and is a very handsome sterling silver cup, valued at $125.)

The Executive Committee decided to offer this cup at the National Dairy Show this year, to be awarded to the best Young Herd, bred and owned by the exhibitor, except the bull, which may be bred by some other breeder in the United States. The animals comprising the herd shall be registered in the Ayrshire Record, and the exhibitor shall be a member of the Association, and entries shall be confined to the United States. The number of animals in the herd and the age of the animals shall follow the classification of the National Dairy Show.

THE PRESIDENT—I wish in behalf of the Association to thank Mr. Stickney for his generous gift.

THE SECRETARY—The matter of advertising the Association in the newspapers should be discussed and settled at this meeting..

THE PRESIDENT—We have some representatives of papers in which we have been accustomed to advertise, and perhaps they would like to say a word.

Mr. Gilbert—How much money was expended last year.

The President—$200 for continuous advertisements in two papers, *The Country Gentleman* and *Hoard's Dairyman;* also some $30 in advertising the Year Book for short periods.

Mr. Wm. T. Wells—Is the advertising left to the discretion of the Secretary?

The President—Until last year it was decided definitely at the annual meetings.

Mr. Gilbert—I move we appropriate $300 for advertising.

Mr. Dodge—I think the appropriation is right, and if we are going to push the Ayrshires we must advertise. Also I think a change of form attracts attention more than a uniform advertisement running the whole year without change. I think we should decide what papers to patronize, for we have papers that will print anything we ask of them that will benefit the breed, and I believe such papers should have the preference over those who do not do it. In other words, help those who help us.

Mr. Burkett—I am not a member of this Association, but if it is not inconsistent, I would like to say a word. I am an editor of a paper that has a very large circulation. We have been doing a great deal for the Ayrshire Association, but we have never gotten anything out of the Ayrshire breeders. If you do not care to advertise in the *American Agriculturist,* for Heaven's sake don't advertise in it, but if you wish to do something for the Ayrshire breed, advertise in the *American Agriculturist.* We publish things of considerable interest to the Ayrshire breeders and to other people. We are trying right now to stir up an interest in live stock. I believe in pure bred stock, but I believe if you want the help of the *American Agriculturist* you can't expect any support from them if you give your advertising to the other papers. If the breeders will send these items in we will be very glad to print them in order to promote the interest in the breed. The agricultural papers have got to live like anything else.

Mr. Gilbert—I would repeat my motion that we appropriate $300 for advertising, and that the disposal of the money be left with the Secretary.

Mr. Oakey—It is all right to leave the disposal of the money in the hands of the Secretary, but I think we should decide at this meeting what papers we wish to patronize.

Mr. Wm. T. Wells—If the Secretary has any serious objections, I think he should state them.

The Secretary—I think it much better for the Association to decide here in open meeting just what they wish to expend in advertising, and direct what papers they wish advertisements inserted in. You appropriate $300, to be divided among a lot of good papers. Every agricultural paper will tell you their paper is the best medium for advertising the breed of any in the United States, and will give good reasons for their statement. What am I to do with $300 and as many good papers wanting the whole sum? I decidedly wish the meeting would settle just what they wish me to do with the $300, for I could use $3,000 to good advantage.

Mr. Converse—I would like to amend the resolution by allowing the Secretary to run an advertisement for a year, but not to spend more than $100 on any one paper. A much better bargain can be made for a year than for a shorter time.

The amendment was carried and the matter of placing the $300 in the hands of the Secretary was passed.

Mr. Yeaton—There is a matter which interests me more than anyone else, perhaps, in relation to the milk and butter record of Lady Fox 9669, which was made before the present rules for advanced registry were adopted. The record was partly private and partly official, as much official as the rules of that day required, but not in conformity with the rules of to-day, and the question is, could the Association, by a vote, order her entered in the Advanced Registry List?

Mr. Spalding—Will you not be establishing a dangerous precedent by opening the door for semi-official records, made before the present rules were adopted? Where can you draw the line after you have admitted one?

Mr. Schanck—I would move that Mr. Yeaton prepare an account of the dairy performances of Lady Fox and send it to the Secretary, who shall be instructed to publish it in the forthcoming Year Book, together with her picture.

The motion was carried.

MILK AND BUTTER RECORD OF LADY FOX 9669

As Prepared by Her Owner, Geo. H. Yeaton.

In the year 1895 Lady Fox 9669 gave 10,036 pounds of 4.70% milk, making 550 pounds of butter.

In 1897 she gave 10,076 pounds of milk, averaging 4.38%, amounting to 507 pounds of butter.

The total amount produced for the two consecutive years was 20,112 pounds of milk and 1,057 pounds of butter.

In the twelve months from April 28, 1896, to April 28, 1897, she produced 12,299 pounds of milk.

She was in the "Home Dairy Test" for the year 1896, and was tested two days in June and two days in December, according to the rules of the Association, by a representative of the Experiment Station of New Hampshire, when her milk had an official average test of 4.30% of butter fat, and was so reported in the annual meeting for that year.

The milk from Lady Fox for the year, amounting to 12,299 pounds, testing 4.30%, makes her butter record 617 pounds for the year.

This record of Lady Fox was made under normal conditions, when she was having nothing more than ordinary care, and without any extra feed, being tied at her stanchion or turned to the pasture a mile from the stable, with the other members of the herd.

The record was made and Lady Fox passed on to the "green pastures and beside the still waters" that are prepared for such as her, long before the rules for "Advanced Registry" were formulated; hence her name does not appear in the list of Advanced Registry Ayrshire Cattle.

It was resolved that the sense of the meeting was that the National Dairy Show should be held in New York in the fall of 1909, and the following committee was appointed to use their best endeavor to have the National Dairy Show come to New York City:

W. P. Schanck of New York.

John W. Oakey of Pennsylvania.

M. B. Gilbert of Massachusetts.

It was voted to request Fair Associations to count ages of cattle entered for exhibition from August 1.

Mr. Spalding—I would like to introduce the following resolution:

WHEREAS, Geo. H. McFadden, L. A. Reymann, Percival Roberts, Jr., and John R. Valentine each contributed $100 as an inducement for the breeders to raise the records of the Ayrshire cow; and

WHEREAS, The breeders of the United States have bred and developed the four champion Ayrshires of the world, thereby doing the Ayrshire breed a great benefit; therefore,

Resolved, As a recognition of their generosity and good will a vote of thanks be extended to the gentlemen, and that a copy of these resolutions be sent to each.

The resolution was unanimously adopted.

The meeting adjourned to meet in the banquet room, where about 50 sat down to a fine spread, which was followed by a very enjoyable hour of speeches by those present.

EXTRACTS FROM PRESIDENT'S ADDRESS

Delivered at the Annual Meeting of the Canadian Ayrshire Breeders' Association at Montreal, 1909, by R. R. Ness.

The more I see of our country, the stronger becomes my conviction that the Ayrshire is the "farmer's cow." Possibly we have not been so progressive as the lovers of some of the other dairy breeds, nor have we advertised and boomed with great acclaim "Our Favorites" as they have done. Here, mayhap, we can take a lesson from the advocates of our sister dairy breeds. We are living in the most progressive age the world has ever known, an age, too, of great opportunities, and right here in this Canada of ours, where the dairy industry is paramount, we Ayrshire breeders have splendid opportunities. Our desire should be to become more capable men. We need to educate ourselves more thoroughly in our profession, become more thoroughly acquainted with Nature's laws and conditions of breeding, to familiarize ourselves with the true Ayrshire type, and to be more careful in selecting sires with which to mate our females, with a view to combine and maintain type, symmetry and vigor of constitution with UTILITY. The former without utility is valueless. Utility without a corresponding type, symmetry and vigor of constitution, enhances the value of our registered stock but little. The two combined add to the value to a very material extent. Herein is a strong point in favor of the Ayrshire. No breed of dairy cattle so possesses and combines these qualities, making them beautiful to look upon and profitable in the dairy.

This also applies to the show ring. As it has been in the past, so will it be in the future. Type, symmetry and vigor of constitution, combined with high milking qualities, must be the stamp of cow that will win under the critical eye of the expert judge. While there may be differences of opinion on minor points, yet this must be the fundamental principle on which awards are made where no dairy test exists. In the dairy test a true estimate of the butter fat and solids demonstrates the ability of the cow or heifer to assimilate food and economically convert it into milk. No dairy test is complete unless the cost of production (amount of feed consumed) is taken into consideration, not only during the test, but for at least two days previous to the commencement of the test, when it extends for only 48 or 72 hours.

The Record of Performance test is the most valuable, not only to breeders of registered dairy cattle, but to our dairymen as a whole. It enables the breeder to determine the true value of his herd and the dairyman to select his sire from a family known to be producers.

While we have had a large number of entries to this test since its commencement, yet we should have more. Ayrshire breeders, awake to your opportunity. Enter your cows in this test and show to the world that the AYRSHIRE cow is "very much alive" in making large records of milk and fat.

When in Scotland last spring, I found the Scotchmen paying greater attention to milk records. It will not be long before the Scotch Ayrshire will have a milk and butter record attached to her pedigree. This is a step forward. It enhances the value of a dairy animal.

In Scotland, the United States and Canada during the past year, in a 365-day test, the Ayrshire cow has demonstrated to the public that she is capable of making large records of milk and fat, as from 10,000 to 13,000 pounds of milk and from 400 to 500 pounds of butter in one season is not uncommon. In the United States over 15,000 pounds of milk and 700 pounds of butter has been produced by one cow in a year. In our Canadian test we have cows that have made records of over 13,000 pounds of milk and 566 pounds of butter. It may be only a short time when we will discover that in the Ayrshire breed we have cows with a milk capacity of 15 or 20 times their own weight within the 12 months.

EXTRACTS FROM THE REPORT OF THE SECRETARY OF THE CANADIAN AYRSHIRE BREEDERS' ASSOCIATION
read at their annual meeting at Montreal in 1909, by W. F. Stephen:

In presenting my third annual report as your Secretary, it is gratifying to note that success has attended the labors of the Canadian Ayrshire breeders. Despite the financial depression and short crops, sales have been numerous and at remunerative prices. More of our dairymen each year are seeking for superior sires. The Ayrshire is becoming noted as a hardy, vigorous animal, capable of high production, either on the bleak uplands or the fertile valleys, under summer skies or in winter conditions. In whatever clime it may be her lot to live, the Ayrshire cow remains the same persistent milker, and is always ready to perpetuate in her offspring those characteristics that enable her to excel as the best all-round farmer's cow.

No dairy breed produces milk so economically as the Ayrshire. Her milk, possibly, is the best suited for all conditions. From her milk may be produced the finest cheese or the choicest butter. As a market milk none excels it. It is rich in butter fat and solids not fat,—those constituents which give milk its value as a food,—and is always of high color. Owing to the fat globules being small in size and very firm, it stands transportation admirably.

The advocates of those dairy breeds that do not give milk of as high quality as the Ayrshire have become very wise. They seek to inform the milk consumers that they are throwing away money in buying high class milk of from 3.5 to 4 per cent. fat. They tell us that it is indigestible and one of the chief causes of infant mortality. They would have us believe that milk testing 3 per cent. and less of fat is much suprior as an article of diet, especially for infants and invalids. The consumer is not so easily fooled. The city fellow is going to have the best he can get. This is why the milk from Ayrshire herds is much sought after in all our leading cities.

There are those who will ridicule these statements, but I know whereof I speak, because I have investigated most carefully the conditions governing the production, sale and consumption of milk, not only in this city in which we meet, but in other cities in Canada and the United States.

I do not speak as I have by way of deprecating any other breed. Far be it from me to do so. There is need and room for all the dairy breeds we have in Canada. Should they be perpetuated as quickly as nature will allow, we will not be able to supply the wide areas of this Dominion with registered stock as fast as required to improve the live stock of the country.

I am no prophet, but I have strong convictions that the farmers of Eastern Canada especially, will more and more turn their attention to intensive farming, with dairying as the chief factor. With our cities growing at a rapid pace, a larger supply of milk is required each year. As the demand for good, wholesome, well-balanced milk increases, so will the demand for the Ayrshire cow increase, provided our breeders are alive to their opportunities and ready to meet these conditions. All we ask for the Ayrshire is that she be given a "square deal." She will do her part if we do ours. Fellow breeders, what are you doing to assist in maintaining the position of pre-eminence held by "Our Favorites"? This is a pertinent question that we as breeders must seriously consider. 'Tis not enough that we realize that we have the best breed. We must *prove* that we have by *record* work such as is being done by other dairy breeds. The best test is the Record of Performance, as conducted under the supervision of the Dominion Department of Agriculture.

Exhibitions.—The exhibits of Ayrshires at all the leading fairs and exhibitions excelled those of former years. Particularly may this be said of our Maritime exhibitions, where a marked improvement over former years was noticeable. We congratulate the Ayrshire breeders of these provinces on the success attending their labors.

The Dominion Exhibition was held in Calgary, Alberta, at too great a distance from the Ayrshire breeding ground to permit a large display. In spite of this, Messrs. Robert Hunter & Sons of Maxville, Ont., and R. R. Ness of Howick, Que., went with splendid exhibits at their own expense, to maintain the dignity of the Ayrshire breed. So much were their exhibits appreciated that many residents of that and other western provinces, about to turn their attention to dairying in the near future, decided that the Ayrshire was the breed for that western country. During the circuit of exhibitions taken in by these breeders a number of sales were made. Better still, since returning home, these and many other breeders have received numerous inquiries for stock. Three car lots of Ayrshires have gone to Alberta since November 1, 1908, and inquiries for more have been received. We have now a number of members in the West, one as far north as Fort Vermillion, in the Peace River District. Ayrshires are proving their worth in that part of Canada, which will provide a market for many years to come for the surplus stock of the eastern breeders.

The following are the winnings and records at the Winter Fair Dairy Tests:—

GUELPH, ONT.—(72-Hour Test.)
Cow, 48 Months or Over.

	LBS. MILK.	LBS. FAT.	POINTS.
Rosalie of Hickory Hill—			
N. Dyment, Clappison, Ont.	147.3	5.89	190.63
Sarah 2nd—			

H. & J. McKee, Norwich, Ont.	165.0	5.71	190.16

Victoria—

H. & J. McKee, Norwich, Ont.	151.4	5.60	185.34

Heifer, 36 Months and Under 48.

Star's Nancy—

H. & J. McKee, Norwich, Ont.	122.0	4.80	155.15

Heifer, Under 36 Months.

	LBS. MILK.	LBS. FAT.	POINTS.
Queen Jessie of Springhill—			
H. & J. McKee, Norwich, Ont.	108.3	4.47	144.61
Star's Sarah—			
H. & J. McKee, Norwich, Ont.	105.6	4.23	139.33
Beauty of Hickory Hill—			
N. Dyment, Clappison, Ont.	93.4	3.29	108.97

OTTAWA, ONT.

(72-Hour Test Computed on Old Basis.)

Jean Armour—			
George Rice, Tillsonburg, Ont.	186.3	7.28	206.94
White Floss—			
H. & J. McKee, Norwich, Ont.	143.7	5.98	168.28
Sarah 2nd—			
H. & J. McKee, Norwich, Ont.	155.8	5.44	167.42
Maggie Brown of Hickory Hill—			
N. Dyment, Clappison, Ont.	146.6	5.49	159.94
Spottie—			
H. & J. McKee, Norwich, Ont.	137.3	5.29	154.20
Rosalie of Hickory Hill—			
N. Dyment, Clappison, Ont.	136.4	4.91	146.50

Heifer, Under 42 Months.

Forget-Me-Not—			
N. Dyment, Clappison, Ont.	106.0	4.41	126.23
Star's Sarah—			
H. & J. McKee, Norwich, Ont.	98.1	3.79	118.00

AMHERST, N. S.—(48-Hour Test.)

Cow, 36 Months and Over.

Victoria of Springbrook—
  McIntyre Bros., Sussex, N. B.   80.1   3.28   113.4
Travellers H. Jennie—
  F. S. Black, Amherst, N. S.   71.9   2.76   94.00
Maggie of Springbrook—
  McIntyre Bros., Sussex, N. B.   66.8   2.36   85.9
Pinafore 2nd of Springvale—
  F. S. Black, Amherst, N. S.   73.8   2.37   78.4

Heifer, Under 36 Months.

White Lady of Springbrook—
  McIntyre Bros., Sussex, N. B.   53.9   2.52   87.00
Hazel of Springbrook—
  McIntyre Bros., Sussex, N. B.   48.6   1.98   72.00
Pinafore of Springbrook—
  McIntyre Bros., Sussex, N. B.   49.1   1.80   67.00
Almeda of Springbrook—
  McIntyre Bros., Sussex, N. B.   54.7   1.79   65.00

(These heifers were all well on in lacteation.)

Record of Performance Test.—A large number of cows and heifers are competing in this test. Although the past two years have been unfavorable for making a large test owing to short pasturage and scant feed, yet some splendid records have been made. Altogether 41 cows and heifers have qualified, 14 since the last report was issued. Their records are as follows:

Mature Class.

                     LBS.   LBS.
                     MILK.   FAT.   DAYS.

Daisy Queen—9705—
  E. Cahoon, Harrietsville, Ont.   13,158   485.39   365

Trixy—9707—

	LBS. MILK	LBS. FAT	DAYS
E. Cahoon, Harrietsville, Ont.	11,222	446.26	365

Dollie Dutton of St. Anne—10005—

A. C. Wells & Son, Sardis, B. C.	10,424	442.70	350

Kirsty 2nd of Neidpath—10125—

W. W. Ballantyne, Stratford, Ont.	9,521	381.95	344

Three Year Old Class.

	LBS. MILK	LBS. FAT	DAYS

Woodroffe Lady Nancy—21454—

A. C. Wells & Son, Sardis, B. C.	7,197	303.91	302

Two Year Old Class.

Adalia 3rd—22948—

E. Cahoon, Harrietsville, Ont.	8,845	326.46	365

Daisy Queen 2nd—22950—

E. Cahoon, Harrietsville, Ont.	6,644	250.18	345

Lady Brant of Neidpath—21463—

W. W. Ballantyne, Stratford, Ont.	6,631	303.99	319

Ruby Royal of the Hills—23375—

A. C. Wells & Son, Sardis, B. C.	6,515	276.45	365

Dollie Dutton of St. Anne 2nd—23374—

A. C. Wells & Son, Sardis, B. C.	6,290	287.72	334

Rosebud—22305—

Joseph Thompson, Sardis, B. C.	7,982	280.10	365

Isaleigh Claribella—23712—

J. N. Greenshields, Danville, Que.	8,454	322.55	365

Annie of Warkworth—21493—

A. Hume & Co., Menie, Ont.	6,689	284.49	365

Stadacona Silver Queen—20043—

G. A. Langelier, Cap Rouge, Que.	6,375	303.38	340

Quite a few cows and heifers went over the maximum of fat required to qualify, but did not give the required amount of milk. Others gave the required amount of milk and fat, but did not freshen within 15 months from the commencement of the test. I think that when a cow makes a good record and does not freshen within the time required, her name and record should be published, but no Certificate issued.

You will notice that the cow Daisy Queen produced 13,158 pounds of milk and 485.39 pounds of butter fat, equivalent to 566.28 pounds of butter. This is the highest known official record made by a Canadian Ayrshire. The highest known official record of an Ayrshire is that made by the cow Rena Ross 14539, owned by John R. Valentine, Bryn Mawr, Pa., U. S. A. It is 15,073 pounds of milk and 751 pounds of butter.

UNIFORM SCALE OF POINTS

As Suggested by a Joint Committee from the United States and Canada Ayrshire Breeders' Association, 1906.

Scale of Points for Ayrshire Bull.

Head	16
Forehead—Broad and clearly defined	2
Horn—Strong at base, set wide apart inclining upward	1
Face—Of medium length, clean cut, showing facial veins	2
Muzzle—Broad and strong without coarseness	1
Nostrils—Large and open	2
Jaws—Wide at the base and strong	1
Eyes—Moderately large, full and bright	3
Ears—Of medium size and fine, carried alert	1
Expression—Full of vigor, resolution and masculinity	3

NECK—Of medium length, somewhat arched, large and strong in the muscles on top, inclined to flatness on sides, enlarging symmetrically towards the shoulders, throat clean and free from loose skin...................... 10

FOREQUARTERS 15
    Shoulders—Strong, smoothly blending into body, with good distance through from point to point and fine on top.......... 3
    Chest—Low, deep and full between and back of forelegs 8
    Brisket—Deep, not too prominent and with very little dewlap...................... 2
    Legs and Feet—Legs well apart, straight and short, shanks fine and smooth, joints firm, feet of medium size, round, solid and deep 2

BODY 18
    Back—Short and straight, chine strongly developed and open-jointed............. 5
    Loin—Broad, strong and level............ 4
    Ribs—Long, broad, strong, well sprung and wide apart 4
    Abdomen—Large and deep, trimly held up with muscular development............. 4
    Flank—Thin and arching................. 1

HINDQUARTERS 16
    Rump—Level, long from hooks to pin bones 5
    Hooks—Medium distance apart, proportionately narrower than in female, not rising above the level of the back.............. 2
    Pin Bones—High, wide apart............. 2
    Thighs—Thin, long and wide apart........ 4
    Tail—Fine, long and set on a level with back 1

Legs and Feet—Legs straight, set well apart, shanks fine and smooth; feet medium size, round, solid and deep, not to cross in walking 2

SCROTUM—Well developed and strongly carried.. 3
Rudimentaries, Veins, etc. Teats of uniform size, squarely placed, wide apart and free from scrotum; veins long, large, tortuous, with extensions entering large orifices; escutcheon pronounced and covering a large surface 4

COLOR—Red of any shade, brown or these with white, mahogany and white, or white; each color distinctly defined................. 3

COVERING 6
Skin—Medium thickness, mellow and elastic 3
Hair—Soft and fine 2
Secretions—Oily, of rich brown or yellow color 1

STYLE—Active, vigorous, showing strong masculine character, temperament inclined to nervousness, but not irritable or vicious........ 5

WEIGHT at maturity not less than 1,500 pounds.. 4

Total 100

SCALE OF POINTS FOR AYRSHIRE COW.

HEAD 10
Forehead—Broad and clearly defined...... 1
Horns—Wide set on and inclining upward.. 1
Face—Of medium length, slightly dished, clean cut, showing veins................ 2

Muzzle—Broad and strong without coarseness, nostrils large 1
Jaws—Wide at the base and strong........ 1
Eyes—Full and bright, with placid expression 3
Ears—Of medium size and fine, carried alert 1

NECK—Fine throughout, throat clean, neatly jointed to head and shoulders, of good length, moderately thin, nearly free from loose skin, elegant in bearing 3

FOREQUARTERS 10
Shoulders—Light, good distance through from point to point but sharp at withers, smoothly blending into body........... 2
Chest—Low, deep and full between and back of forelegs 6
Brisket—Light 1
Legs and Feet—Legs straight and short, well apart, shanks fine and smooth, joints firm; feet medium size, round, solid and deep 1

BODY 13
Back—Strong and straight, chine lean, sharp and open-jointed 4
Loin—Broad, strong and level............. 2
Ribs—Long, broad, wide apart and well sprung 3
Abdomen—Capacious, deep, firmly held up with strong muscular development....... 3
Flank—Thin and arching................. 1

HINDQUARTERS 11
Rump—Wide, level and long from hooks to

pin bones, a reasonable pelvic arch
allowed 3
Hooks—Wide apart and not projecting above
back nor unduly overlaid with fat........ 2
Pin Bones—High and wide apart.......... 1
Thighs—Thin, long and wide apart........ 2
Tail—Long, fine, set on a level with the
back 1
Legs and Feet—Legs strong, short, straight
when viewed from behind and set well
apart; shanks fine and smooth, joints firm;
feet medium size, round, solid and deep.. 2

UDDER—Long, wide, deep but not pendulous,
nor fleshy; firmly attached to the body, extending well up behind and far forward;
quarters even; sole nearly level and not indented between teats, udder veins well
developed and plainly visible................. 22

TEATS—Evenly placed, distance apart from
side to side equal to half the breadth of udder, from back to front equal to one-third
the length; length 2½ to 3½ inches, thickness in keeping with length, hanging perpendicular and not tapering..................... 8

MAMMARY VEINS—Large, long, tortuous
branching and entering large orifices........... 5

ESCUTCHEON—Distinctly defined, spreading
over thighs and extending well upward......... 2

COLOR—Red of any shade, brown or these
with white; mahogany and white, or white;
each color distinctly defined . (Brindle markings allowed but not desirable)................ 2

COVERING 6
    Skin—Of medium thickness, mellow and elastic 3
    Hair—Soft and fine...................... 2
    Secretions—Oily, of rich brown or yellow color 1

STYLE—Alert, vigorous, showing strong character; temperament inclined to nervousness but still docile 4

WEIGHT at maturity not less than 1,000 pounds 4

    Total 100

CHARTER.

AN ACT TO INCORPORATE THE AYRSHIRE BREEDERS' ASSOCIATION.

It is hereby enacted by the General Assembly of the State of Vermont:

SEC. 1. J. D. W. French, James F. Converse, Alonzo Libby, F. H. Mason, Obadiah Brown, Henry E. Smith, C. M. Winslow, S. M. Wells, H. R. C. Watson, James Scott, George A. Fletcher, Charles H. Hayes, John Stewart, their associates and successors, are constituted a body corporate by the name of the "Ayrshire Breeders' Association," and by that name may sue and be sued; may acquire by gift or purchase, hold and convey real and personal estate necessary for the purposes of this corporation, not to exceed twenty-five thousand dollars; may have a common seal and alter the same at pleasure.

SEC. 2. The object of this corporation shall be to publish a Herd Book, and for such other purposes as may be conducive to the interests of breeders of Ayrshire cattle.

SEC. 3. This corporation may elect officers and make such by-laws, rules and regulations for the management of its business as may be necessary, not inconsistent with the laws of this State.

SEC. 4. This corporation may hold its meetings at such time and place as the corporation may appoint.

SEC. 5. This act shall take effect from its passage.

JOSIAH GROUT,
Speaker of the House of Representatives.

LEVI K. FULLER,
President of the Senate.

Approved November 23, 1886.

EBENEZER J. ORMSBEE,
Governor.

(A true copy.)

Attest: E. W. J. HAWKINS, Engrossing Clerk.

CONSTITUTION.

Preamble.

We, the undersigned breeders of Ayrshire cattle, recognizing the importance of a trustworthy Herd Book that shall be accepted as a final authority in all questions of pedigree, and desiring to secure the co-operation of all who feel an interest in preserving the purity of this stock, do hereby agree to form an Association for the publication of a Herd Book, and for such other purposes as may be conducive to the interests of breeders, and adopt the following Constitution:

Article I.

This Association shall be called the Ayrshire Breeders' Association.

Article II.

The members of the Association shall comprise only the original signers of this Constitution, and such other persons as may be admitted, as hereinafter provided.

Article III.

Sec. 1. The officers of the Association shall consist of a President, four Vice-Presidents, a Treasurer, a Secretary and an Auditor, who together with six members of the Association, all chosen by ballot, shall constitute an Executive Committee.

Sec. 2. The President, Secretary and Treasurer shall be the Finance Committee ex officio.

Sec. 3. The President, Vice-Presidents, Treasurer, Secretary and Auditor shall be elected annually.

The six members who make up the balance of the Executive Committee shall be elected as follows: Two members for one year, two members for two years and two members for three years, and hereafter two members shall be elected each year for a term of three years.

SEC. 4. The President shall preside at all meetings of the members of the Association, and all meetings of the Executive Committee when he is present, but when absent a Vice-President shall act in his stead. The President shall sign all Certificates of Membership which may be issued, and shall be the custodian of all bonds given by officers of the Association, or renewals thereof.

SEC. 5. The Finance Committee shall have authority to take the entire control and management of the affairs of the Association, between the Annual Meetings, with full power and authority to do what they deem proper and best for its interests, but nothing contrary to the expressed wish of the Association.

SEC. 6. The Treasurer shall have charge of all the funds of the Association and make all investments thereof, subject to the provisions of the section regulating the Finance Committee, and shall pay all bills of the Association, after being indorsed by the Finance Committee and approved by the Auditor, and shall perform such other duties as are incident to the office of Treasurer. He shall give a bond with sureties, to the satisfaction of the Finance Committee and Auditor.

SEC. 7. The Secretary shall be the corresponding and recording officer of the Association, shall sign and issue all certificates of membership and registry and of transfer registry, and shall keep a record of all such certificates issued, and do such other duties as are incident to the office of Secretary.

He shall edit and publish the Herd Book at such times and in such form as the Executive Committee may direct.

He is authorized to expend such sums as he may find necessary for the carrying on the ordinary business of his office, and shall keep an accurate account in detail of all moneys received and paid out by him in the performance of his duties, a copy of which he shall transmit quarterly, during the week next succeeding the quarter, to the Auditor, and shall at the same time send to the Treasurer whatever moneys he may have on hand at the ending of the quarter.

He shall give a bond with sureties to the satisfaction of the Finance Committee and Auditor.

SEC. 8. The Finance Committee shall annually examine the condition of the Association in its financial and business affairs, and report its condition to the Association at its Annual Meeting, and in conjunction with the Treasurer shall act in making investments of the funds of the Association.

Any disagreement between the Finance Committee as to the investment or care of the funds of the Association shall be referred to the Executive Committee for final adjustment.

All bills against the Association shall be approved by the Finance Committee and sent by them to the Auditor.

SEC. 9. The Auditor shall examine all accounts sent him from any member of the Finance Committee, and if found correct, shall approve and forward the same to the Treasurer for payment, and shall annually, when auditing the accounts of the year for the Secretary and Treasurer, previous to the Annual Meeting, make a complete inventory of all property found in the hands of the Secretary and Treasurer, and forward the same to the Finance Committee, which shall be incorporated in the report of the Finance Committee to the Association at their Annual Meeting.

SEC. 10. The Treasurer, Secretary and Auditor shall receive such compensation for their services as the Association shall determine.

ARTICLE IV.

The Annual Meeting of the Association shall be held each year at such time and place as shall be designated by the Executive Committee (of which notice shall be sent to members at least one month previous) for the discussion of questions of interest to the members, and for the election of officers for the ensuing year. Special meetings of the Association may be called by the President or by the Executive Committee, or at the written request of ten members. Twenty days' notice must be given and the object of the meeting announced in the call, and no business other than that specified in the call shall be transacted at the special meeting. Time and place shall be determined in same way as Annual Meeting.

At all meetings of the Association members may vote in person or by proxy, or they may send their ballot by mail to the Secretary, whose duty it shall be to vote the same, and to acknowledge their receipt. At least twenty members present, represented by proxy or written ballot, shall be a quorum for transacting business.

ARTICLE V.

Only breeders of Ayrshire cattle shall be eligible for membership, and members shall be elected at any regular meeting of the Association; also by the unanimous written consent of the Executive Committee at any time between the annual meetings, subject to the following conditions:

Each applicant for membership shall be recommended by one or more members of the Association as a trustworthy and careful breeder; and no new member shall be admitted if objected to by any officer of the Association.

The Secretary shall notify the candidate of his rejection, or, in case of his election, that he will be admitted as a member on signing the Constitution and paying the initiation fee.

An applicant who has been rejected shall not be voted on again until two years from the date of his rejection, unless by the unanimous consent of the officers of the Association.

Article VI.

Each member shall pay an initiation fee of twenty-five dollars. These fees shall constitute an Association fund to defray the expenses of publishing the Herd Book, and other charges incidental to the organization of the Association, and to the transaction of its business.

No officer or member shall be authorized to contract any debt in the name of the Association.

Article VII.

The Herd Book shall be edited by an editor appointed for that purpose under the control and supervision of the Executive Committee, and shall be published only with its official approval.

The charge for entry of the pedigree of each animal belonging to a member of the Association shall be fixed by the Executive Committee, but shall not exceed one dollar, except for an animal two years old.

Animals not belonging to members of the Association may be entered in the Herd Book upon the payment of twice the amount charged to members.

The Herd Book charges shall be appropriated to the examination and verification of pedigrees and the preparation of the Herd Book, which shall be published by the Association and be its property. The price of the Herd Book shall be determined by the Executive Committee. The editor shall keep on file all documents constituting his authority for pedigrees, and shall hold them subject to the inspection of any member of the Association, and shall deliver them to his successor in office.

Article VIII.

Should it occur at any time that any member of the Association shall be charged with wilful misrepresentation in regard to any animal, or with any other act derogatory to the standing of the Association, the Executive Committee shall examine into the matter; and, if it shall find there is foundation for such a charge, the offending member may be expelled by a vote of two-thirds of the members of the Association present or represented at any regular meeting.

Article IX.

This Constitution may be altered or amended by a vote of two-thirds of the members present or represented by proxy at any annual meeting of the Association.

Notice of proposed alterations or amendments shall be given in the call for said meeting.

REGULATIONS.

1. Only such animals shall be admitted to the Herd Book as are proved either to be imported from Scotland, or descended from such imported animals.

2. All animals hereafter imported to be eligible to registry in the Ayrshire Record must previously be recorded in the Ayrshire Herd Book of Scotland, and an application for registry must be accompanied by a cer-

tificate of registry duly signed by the Secretary in Scotland.

Entries of calves imported in dam must be accompanied by the certificate of registry of sire and dam in the Scotch Herd Book, also certificate of bull service signed by owner of bull.

3. No animal not already named and entered in some Herd Book shall be accepted for entry under a name that has already been offered for entry; also, the affix 1st, 2d and 3d shall apply only to calves of the cow bearing the name used; not to her grandchildren or any other animal.

4. The breeder of an animal shall be considered the one owning the dam at the time of her service by the bull.

5. No pedigree will be received for entry from any one, except the breeder of the animal offered, unless it is accompanied by a certificate of the breeder or his legal representative, indorsing the pedigree.

Entries of calves, sired by bulls not owned by the breeder of the calf, shall be accompanied by a certificate of bull service signed by owner of bull.

6. All animals sold, in order that their progeny may be registered, must have their successive transfers duly recorded. Records of transfers will be made only on the certificate of former owner, or his legal representative.

7. A transfer-book shall be kept by the Editor, in which all changes of ownership shall be recorded.

8. The Editor shall keep a record of the deaths of all animals which may be sent to him. (And breeders are requested to forward the same, stating cause, etc.)

9. The fees for recording are one dollar for each animal recorded by and in the name of a member of the

Association, being either bred or owned by him, and two dollars for animals over two years old at the time of entry, but this is not intended to allow at members' rates, the recording of calves born after the dam is sold, when the owner is not a member.

On imported animals the two-year limit is reckoned from the date of importation, and the same on animals brought from Canada.

A fee of twenty-five cents will be charged for recording ancestors necessary to complete a pedigree to importation or to cattle already in the Ayrshire Record, when the record is for cattle bred and owned by other parties, and is of no other value to the person having the recording done, other than to admit his animal to record.

Transfer fee, twenty-five cents.

Double the above rates are charged to those not members.

Duplicate certificates of entry or transfer, twenty-five cents each.

A fee of fifty cents will be charged for a Custom House certificate on each animal imported from Canada.

All the above fees should accompany the entry or transfer papers to insure attention.

10. An individual membership shall be continued after the death of a member in the settlement of his estate until the same shall be settled, and then the membership shall cease. The inheritor of a herd of Ayrshires shall also inherit the membership of the Ayrshire Breeders' Association—subject to approval of said Association. In case of corporations, the corporation may continue as a member so long as they are interested in the Association, and shall be represented by such person as may be designated by the President and Secretary of the corporation.

The surviving member of a firm may be the member of the Association.

A firm shall have but one address.

11. These Regulations may be altered, amended or added to, with the consent of two-thirds of the officers of the Association and Executive Committee.

GENERAL INFORMATION.

Each volume of the Ayrshire Record, I to XIX, inclusive, may be obtained of the Secretary; postage paid, $2.25.

Milk record blanks to accommodate herds of thirty-three cows may be had of the Secretary, C. M. Winslow, Brandon, Vt., $1.50 per 100.

Blanks for extending pedigrees to five generations may be had of the Secretary at $1.00 per 100, postage paid.

Private Herd Book records, board cover, may be had of the Secretary at $1.50 each, postage paid, arranged for tabulated pedigree for seventy-five (75) cows, with spaces for monthly milk and butter record for eight years, service and produce record for twelve years.

All blanks necessary for recording and transferring Ayrshires may be had of the Secretary free of charge.

Membership fee $25.00, which is for life, not transferable, and no assessments.

The survivor of a partnership may become a member.

The inheritor of a herd may also inherit the membership.

The partnership of a herd can apply to only one herd and cannot be divided for two herds or in two post-office addresses.

Members' fees for recording, $1.00 for each animal under two years old, $2.00 for each animal over two years old.

The date of the two-year limit in age is reckoned from the date the application for record is mailed.

The two-year limit on animals imported or brought from Canada is reckoned from date of Custom House receipt.

Transfer fee, twenty-five cents.

A fee of twenty-five cents each is charged for recording ancestors necessary to complete a pedigree to importation, or to cattle already recorded in the Ayrshire Record when the record is for cattle bred and owned by other parties and is of no other value to the person recording.

Double the above rates to non-members.

The rate charged is governed by the fact of whether the person sending the application is a member or non-member, and not by who bred the animal.

Duplicate certificates of entry or transfer, twenty-five cents each.

A fee of fifty cents each is charged for a Custom House certificate for animals imported from Canada.

Application for entry of Canadian bred animals owned by non-members of the Ayrshire Breeders' Association will have to be accompanied with a Canadian certificate of entry as authority for accepting the pedigree.

No animal will be received for record that does not trace in each branch of its pedigree step by step by name and number to a reliable importation.

All the above fees should accompany the applications to insure attention.

In giving sire and dam be careful to always give the Herd Book number of sire and dam.

When purchasing an animal be sure to get a transfer or see that the seller sends one to the Secretary for record.

When buying a female in calf be sure to get a certificate of bull service from the owner of the cow, and attach it to the application for entry of her calf when sending in for record.

In filling out an application for entry of an animal that is sold there is no need of a separate transfer, but enter it in the line for owner with date of sale, and there is no extra charge for a transfer when so recorded.

In giving the markings on the back of the application blank be careful to mark with ink and as accurately as possible, marking *r* for the red spots and *w* for the white spots.

When buying Ayrshires in Canada our Government admits free of duty if they are registerd in our book before being entered at the Custom House, and it is much safer not to move them from the owner until they are recorded, because sometimes it happens that the pedigree must be looked up at the Canada office, and there is often considerable delay.

It would be wise for any one having an Ayrshire cow of extraordinary dairy ability to have her tested for Advanced Registry.

RULES FOR ADVANCED REGISTRY.

PREAMBLE.

For the purpose of encouraging a better system of keeping milk and butter records, and that we may obtain more reliable records of the dairy yield of Ayrshire cows, we hereby adopt the following rules and regulations for the establishment of a system of Advanced Registry for Ayrshire cattle:

RULE I.

The Secretary of the Association shall have charge of the registry under the general supervision and direction of the Executive Committee, shall prepare and publish blank forms and circulars needed in carrying this system into effect, receive and attend to all applications for this registry, and have general oversight and direction of all official tests of all milk and butter production for it, and perform such other duties as may be required to secure the efficiency and success of this system. He shall make a full report of his work in this branch at the annual meeting each year, and publish the entries when so ordered by the Executive Committee.

RULE II.

All tests shall be for a period of 365 consecutive days.

RULE III.

CLASSIFICATION OF ANIMALS.

Cows from two to three years old shall be in a class known as the two-year-old form.

Cows from three to four years old shall be in a class known as the three-year-old form.

Cows from four to five years old shall be in a class known as the four-year-old form.

Cows above five years old shall be in a class known as the full-age form.

RULE IV.

ELIGIBILITY OF BULLS.

No bull shall be eligible to Advanced Registry unless he shall have been previously recorded in the Ayrshire Record.

a A bull to be eligible to Advanced Registry shall be a typical Ayrshire bull in general appearance, shall scale 80 points and have two daughters in the Registry from different dams.

b. A bull may be admitted to Advanced Registry without physical qualifications and without scaling, provided he has four daughters in the Advanced Registry from different dams.

RULE V.

ELIGIBILITY OF COWS.

No cow shall be admitted to Advanced Registry unless she shall have been previously recorded in the Ayrshire Record.

TWO-YEAR-OLD FORM.

Year's record. If her record begins the day she is two years old, or before that time, she shall, to entitle her to record, give not less than 6,000 pounds of milk in 365 consecutive days from the beginning of the test and 214.3 pounds of butter fat, and for each day she is over two years old at time of beginning the test there shall be added 1.37 pounds of milk to the 6,000 pounds and .06 pounds of butter fat to the 214.3 pounds.

THREE-YEAR-OLD FORM.

If her record begins the day she is three years old, she shall, to entitle her to record, give not less than 6,500 pounds of milk in 365 consecutive days from the beginning of the test and 236 pounds of butter fat, and for each day she is over three years old at the time of beginning the test there shall be added 2.74 pounds of milk to the 6,500 pounds and .12 pounds of butter fat to the 236 pounds, which addition shall be made in each succeeding form to maturity.

FOUR-YEAR-OLD FORM.

Year's record—7,500 pounds of milk and 279 pounds of butter fat.

MATURE FORM.

Year's record—8,500 pounds of milk and 322 pounds of butter fat.

RULE VI.

PERIOD FOR MAKING TESTS.

All tests for a year shall be commenced as soon after calving as practicable, and shall not extend beyond 365 days from the commencement of the test, and in no case shall the test include the milk or butter fat from a second calving.

RULE VII.

APPLICATION FOR TESTS.

An application for a test will not be accepted from a person who is not a member of the Ayrshire Breeders' Association. Application for intended tests should be made to the Secretary as long as possible before the desired time for beginning such tests in order to allow sufficient time to arrange with the Experiment Station of the State where the owner is located for their supervision of the test.

In making application for a test, the owner should give sufficient evidence of the capability of the cow to qualify to warrant making the test.

RULE VIII.

METHOD OF CONDUCTING TESTS.

All tests shall be under the supervision of the Secretary and the Experiment Station of the State where the test is being made, or such persons as may be appointed by concurrence of Secretary and Station.

The owner shall weigh each milking from each cow being tested and keep a careful record of the same on blank forms furnished by the Secretary, and at the end of each month he shall add the amount of milk given by each cow and set it down in the column prepared for that purpose, and send the filled blank to the Secretary as soon as possible after the month is ended.

In addition to this, the Experiment Station will send an agent to the stable each month, to make a two days' test of the milk of each cow in the test, for quantity given at each milking during his visit and for the amount of butter fat in each sample.

He will also inspect the method of weighing and the daily record, compare the weights with those given during his visit and report the same to the Secretary. The milk record kept by the owner of the cows being tested will be accepted as the record for the year, provided it agrees practically with that taken monthly by the agent.

The result of each cow's test shall be computed in the

following manner: The weights of milk produced each month shall be multiplied by the per cent. of butter fat as shown by the official test for that month.

If it is desired to show the amount of commercial butter, it shall be obtained by the Experiment Station method of the addition of one-sixth, being on a basis of 85 per cent. fat.

RULE IX.

All the expense incurred by the Association in the employment of the Experiment Station or their agent in conducting the test shall be divided equally between the Association and the member having the test made. It is expected that the member having the test made will without charge render such assistance as he is able to the agent, in conveyance to the railroad station and in entertainment while making the monthly test.

RULE X.

NO FEE REQUIRED FOR ENTRIES.

In view of the public benefits accruing from investigations under this system of registry, and of personal benefits to owners and breeders of Ayrshire cattle from demonstrations of their superiority by properly authenticated milk and butter records made, gathered and preserved through this system, no fees will be charged for any form of entry in its Register.

RULE XI.

AMENDMENT.

These rules may be altered, amended or added to by a two-thirds vote of the members present at any regular meeting of the Association, notice of proposed amendment having been given in the call for said meeting.

HOME DAIRY TEST 1909-10.

PRIZES OFFERED.

French Prize, Silver Cup	$ 75.00
Cash Prizes	225.00

Two Diplomas.

All of them awarded for milk and butter fat from the uniform scale of points as follows: 1 point for every pound of milk; 17½ points for every pound of fat.

The French Prize of a handsome silver cup is obtained from the income of the French Fund of $1,500, donated by Miss Cornelia A. French, North Andover, Mass., in memory of her brother, the late J. D. W. French.

PRIZES FOR MILK AND BUTTER FAT.

The Ayrshire Breeders' Association offers the following premiums for herds of five cows, and for individual cows, making the best records for one year, under the conditions hereinafter named:

For herds of five cows each, six prizes. First prize, a silver cup valued at $75; second prize, $50; third prize, $40; fourth prize, $30; fifth prize, $20; sixth prize, $10.

For individual cows, six prizes. First prize, $30; second prize, $20; third prize, $15; fourth prize, $10; fifth prize, diploma highly commended; sixth prize, diploma commended.

CONDITIONS OF TESTS.

OPEN TO MEMBERS OF THE AYRSHIRE BREEDERS' ASSOCIATION ONLY.

1. All animals competing must be registered in the Ayrshire Record and stand on the books of the Association as owned by the person competing.

For the year 1909 there will two tests start, one April 1 and one October 1, the object being to change the time of starting the test from April to October, without an interim with no testing. A member of the Association may enter a herd of from five to twelve cows or heifers, test to begin April 1, or he may enter to begin October 1.

A member entering for the test beginning in April may also enter again for the test beginning in October, with the April set entire or in part, or with an entirely new set.

Notice of proposed entry should be given as early as possible, to allow sufficient time to make arrangements with the Experiment Station before beginning the test.

Each contestant shall name from five to twelve cows or heifers to be tested through the year, and at the end of the year he may select any eight of these for the individual cow prizes and for the herd prizes.

At the end of each month each contestant shall report to the Secretary of the Association, upon blanks furnished for such purpose by such office:

a A complete record of weights of each milking, with the correct footing of each for the month.

b The calving and service record for the month.

c An approximate statement of the amount and kind of food given the animals, the manner of stabling and care of same, a full statement for the first month, and after that enter on the blank for that month any change in food or care that may occur from month to month during the year.

These tests shall be under the supervision of the Committee appointed by the Ayrshire Breeders' Association, but any member of the committee owning animals competing in said tests shall be barred from having supervision of his own test, but some other member of the Committee may supervise and take charge of the test. All cows shall be wholly under the control of the owner, so far as feeding and general treatment are concerned.

All tests shall be under the supervision of the Committee and the Experiment Station of the State where the herd being tested is located. In order to save expense and insure careful supervision, arrangements will be made as far as possible to secure the services of Experiment Station agents living not too far from the herd being tested.

The Experiment Station will send an agent monthly to make a two days' test of the milk from each cow in the test, and to make a record of the same on blanks prepared for the purpose, together with the two days previous to the agent's test, as taken from the owner's private record.

The sampling of milk and sending the same to the Experiment Station will be done by the Experiment Station agent.

The daily weights of milk made by the contestant will be accepted, provided they do not materially differ from the weights taken by the agent at his monthly visits.

The result of each year's test shall be computed in the following manner: The weights of milk produced each month shall be multiplied by the per cent. of butter fat as shown by the official test for that month, and the sum of the results thus obtained shall be the year's record for butter fat.

The awards will be made from a uniform scale of points, figuring the total amount of milk from each cow at one point for each pound of milk given during the year, and 17½ points for each pound of butter fat reported by the Experiment Station agent for the year.

The expense incurred from employing the Experiment Station shall be equally divided between the contestant and the Association. It is expected that the contestant will convey the Experiment Station agent to and from railroad station and entertain him free of charge while making the tests.

Testing for Advanced Registry may be made in connection with the Home Dairy Test with no extra expense, and it is advisable to carry along the two at the same time.

>C. M. WINSLOW,
>THOMAS TURNBULL, JR.,
>W. V. PROBASCO,
>Committee on Home Dairy Tests.

OFFICERS OF THE ASSOCIATION.

President.

E. J. Fletcher.....................Greenfield, N. H.

Vice-Presidents.

J. F. Converse.....................Woodville, N. Y.
George E. Pike....................Gouverneur, N. Y.
J. W. Clise...........................Seattle, Wash.
J. A. Ness...........................Auburn, Maine

Secretary and Editor.

C. M. Winslow.......................Brandon, Vt.

Treasurer.

N. S. Winsor.......................Greenville, R. I.

Auditor.

George H. Yeaton.........................Dover, N. H.

Balance of Executive Committee.

W. P. Schanck..............Avon, N. Y., for 3 years
Howard Cook..................Beloit, O., for 3 years
Charles H. Hayes......Portsmouth, N. H., for 2 years
John W. Oakey.........Bryn Mawr, Pa., for 2 years
John R. Valentine.......Bryn Mawr, Pa., for 1 year
L. A. Reymann........Wheeling, W. Va., for 1 year

MEMBERS OF THE ASSOCIATION.

ARIZONA.

McDonald, W. A. ....Mesa

CALIFORNIA.

Bement, George ....Fruitvale

COLORADO.

Osgood, J. C. ....Redstone

CONNECTICUT.

Aiken, Ella R. ....Norwalk
Averill, R. J. ....Washington Depot
Avery, John D. ....North Stonington
Baton, John A. & Son ....Wauregan
Connecticut Agricultural College...............Storrs
Connecticut Insane Asylum...............Middletown
Dorrance, Henry ....Plainfield
Ennis, Alfred A. ....Danielson
Fischer, W. H. ....New Canaan

Greene, B. D. ....Stamford
Kahn, George A. ....Yantic
Larned, J. H. ....Putnam
Manwaring, John ....Norwich
Palmer, Edward G. ....Plainfield
Roode, Joseph ....Jewett City
Sears, N. E. ....Elmwood
Wayside Farm ....Putnam
Weed, John W. ....Noroton
Wells, Dudley & Son....................Wethersfield
Wells, Dudley 2d ....Wethersfield
Wells, William T. ....Newington

ILLINOIS.

Crabb, Frank A. ....Litchfield
Jones, Granville ....Galesburgh
Stewart, John ....Elburn

INDIANA.

Richards, C. C. ....Malotte Park

KANSAS.

Delap, S. N. ....Iola

MAINE.

Bearce, George B. ....Lewiston
Blanchard, S. D. & Sons.....................Sanford
Buckley, J. P. ....Stroudwater
Burnham, M. M. ....Cumberland Centre
Dow, Fred N. ....Portland
Dunn, W. H. & G. H. ....Norway
Good Will Home Association................Hinckley
Hunnewell, A. A. ....New Gloucester
Hunt, H. C. ....Brunswick
McCrum, Lemuel ....Mars Hill
Ness, John A. & Rowland....................Auburn

Pember, Elmer F. ....Bangor
True, Joseph S. ....Lewiston

MARYLAND.

Cashell, F. H. ....Derwood
Harrison, Charles K. ....Pikesville
Scott, J. McPherson ....Hagerstown

MASSACHUSETTS.

Bacon, P. K. ....Campello
Barnes, B. F. ....Haverhill
Beldon, C. M. ....South Natick
Boise, Enos W. ....Blandford
Bowker, George H. ....Westboro
Burt, Jairus F. ....Easthampton
Choate, Charles F. ....Southboro
Cooke, F. C. ....Carlisle
Copeland, Davis & Son.........................Campello
Crissey, Warren ....Great Barrington
Curtis, L. W. ....Globe Village
Doe, Charles C. ....Lexington
Easterbrook Brothers ....Webster
Essex Co. Training School,
   W. Grant Fancher, Supt. ....Lawrence
French, C. A. ....North Andover
Gilbert, M. B. ....Wilmington
Hamilton Woolen Co. ....Southbridge
Harrington, H. A. ....Worcester
Haskell, A. M. ....North Beverly
Knowlton, George W. ....West Upton
Lawrence, James ....Groton
Leach, J. Hooper ....Bridgewater
Leach, Philo ....Bridgewater
Merriam, Herbert ....Weston
Marsh, William H. ....Barre Plains

Morrell, Harry E. ....Wayland
Mt. Hermon Boys' School ....Mt. Hermon
Newton, L. W. ....Ashburnham
Peirce, F. C. ....Concord Junction
Perley, Charles ....Bradford
Pierce, George H. ....Concord
Piper, Anson C. ....South Acton
Ray, Joseph G. ....Unionville
Reed, Hammon ....Lexington
Sage, Charles D. ....North Brookfield
Scott, Thaxter & Son ....Hawley
Smith, Peter D. ....Andover
Stevens, Edmund H. ....Cambridge
Stickney, George E. ....Newburyport
Stone, George F. ....Littleton
Tyler, Arthur F. ....Athol
Walker, William I. ....Great Barrington
Young, Gilman P. ....Grafton

MICHIGAN.
Michigan School for the Deaf ....Flint

MINNESOTA.
Hill, James J. ....St. Paul
Reeve, C. McC. ....Minneapolis
Scott, John W. ....Austin
Wilcox, John F. ....Minneapolis

MISSISSIPPI.
Surget, James ....Natchez

MISSOURI.
University of Missouri....................Columbia

MONTANA.
Davidson, E. M. & Son ....Bozeman

New Hampshire.

Abbott, J. N. ....Concord
Bell, Charles J. ....Hollis
Breck, Stephen R. ....Claremont
Cater, H. F. & Son ....North Barrington
Childs, Harlow N. ....Piermont
Clark, George C. ....Orford
Cross, W. L. ....Ponemah
Doud, Arthur V. ....Bristol
Edes, Samuel ....Newport
Fletcher, Etna J..................South Lyndeboro
Garvin, W. R. ....Dover
Hayes, Charles H. ....Portsmouth
Hayes, Charles S........................Portsmouth
Holt, Andy ....Lyndeboro
Holt, E. A. ....Hudson
Kimball, Herbert M. ....Concord
Marshall, William C. ....Laconia
Rockwood, C. E. & Son......................Temple
Russell, Frank E ....Greenfield
Sawyer, E. E. ....Atkinson
Strafford County Farm......................Dover
The Uplands ....Bridgewater
Upham, Charles H. & Son............Thornton's Ferry
Yeaton, George H. ....Dover

New Jersey.

Beach, Frederick H. ....Morristown
Burke, Joseph F. ....Morristown
Casterline, J. Andrew......................Dover
Crane, John ....Union
Farley, F. C. ....Milburn
Freeman, Charles D. ....Iselin
Glen Alpine Farm ....Morristown
Lindsay, William ....Plainfield

Magie, J. D. & B. P.....................Elizabeth
Probasco, W. V. ....Cream Ridge
Tilton, E. A. ....Hamilton
Whittingham, W. R. ....Milburn

NEW YORK.

Arden Farms Dairy Co., Wm. Viner, Supt......Arden
Ashley, E. L. ....Glens Falls
Atwood, L. E. & Sons.......................Chazy
Babcock, F. M. ....Gouverneur
Ballou, George William.................Middletown
Barnes, N. ....Middle Hope
Barney, C. S. ....Milford
Barney, Kent ....Milford
Bell, George H. ....Rome
Bensley, M. F. ....Buffalo
Bentley, Ellis W........................Windham
Bilby, Emerson ....Deposit
Boyd, Frank H. ....Florida
Brayton, C. N. ....South Wales
Brusie, O. W. ....Millerton
Burdick, George W. ....Friendship
Burdick, Thomas J. & Sons..................Alfred
Button, E. L. ....Melrose
Buttrick, C. A. ....Liberty Falls
Campbell, John S. ....New York Mills
Clark, C. W. ....Guymard
Clark, N. E. ....Potsdam
Colburn, J. L. ....Milford
Conger, Lawton M. ....Collins
Converse, J. F. ....Woodville
Cookingham, F. H. ....Cherry Creek
Crowley, Thomas J. ....Potsdam
Davison, Milton W. ....Camisteo
Delaney, J. J. ....Grindstone

Doane, Franklin ....Middletown
Dorn, Elmer J. ....Johnstown
Dunham, Lawrence.....260 Columbus Ave., New York
Emery, C. G. ....Clayton
Griffin, J. H. ....Moira
Gurnsey, James H. & Co.................... Woodhull
Hall, Lott ....Gouverneur
Ham, Eugene ....Verbank
Hamilton, William Pierson................Sterlington
Harrington, A. D. ....Oxford
Hatch, C. E. ....Gainesville
Heseltine, F. L. & Sons....................West Union
Hill, J. Edwin & Son.....................Gouverneur
Hillman, A. E. & Son........................Cuyler
Horton, H. A. ....Johnson
Howatt, Gerald ....White Plains
Hubbard, George D. ....Camden
Huffstater, L. ....Sandy Creek
Hulett, H. L. ....Allentown
Jackson, B. O. & Son......................Boonville
Jay, William ....Katonah
Jenkins, J. W. ....Vernon
Karr, S. S. & Sons..........................Almond
Kenyon, Louis H. ....Utopia
Lansing, E. Ten Eyck ....Little Falls
Leach, J. S. & Son ....Gouverneur
Lewis, C. W. & Sons...................Alfred Station
Litchard, A. L. & Son......................Rushford
McCrea, Robert ....Champlain
Mercereau, W. W. & H. B. ....Vestal
Nichols, James H. ....Carmel
Norton, W. H. ....Belmont
Oneida Community, Limited ....Kenway
Ormiston, James ....Cuba
Paget, A. H. ....Lakeville

Pike, George E. ....Gouverneur
Ramsdell, H. S. ....Newburgh
Reburn, James ....Sterlington
Rhodes, T. F. ....Camillus
Ricker, Clarence ....Belmont
Rogers, G. L. ....Gouverneur
Ryan, P. ....Brewster
Ryder, Arthur B. ....Barnerville
Schanck, W. P. ....Avon
Schouten, E. A. ....Cortland
Sears, B. C. ....Blooming Grove
Silliman, J. B. ....Kortright
Siver, D. E. ....Cooperstown
Skinner, Harry W. ....Utica
Smith, Oliver & Son.....................Chateaugay
Stetson, Francis Lynde ....Sterlington
Stewart, William H. ....Olean
Story, S. S. ....North Stockholm
Stowell, F. D. & E. ....Black Creek
Stowell, W. C. ....Black Creek
Strickland, J. P. ....Cattaraugus
Taber, George ....East Aurora
Taylor, John L. ....Owego
Tod, William Stewart.........45 Wall St., New York
Topping, R. R. ....Amsterdam
Tubbs, Ambie S. ....Maple View
Tucker, W. G. ....Elm Valley
Tuttle, M. A. ....Hornell
Underhill, C. S. ....Glenham
Verplank, Samuel ....Fishkill-on-Hudson
Ward, C. G. ....Delhi
Ward, M. J. ....Treadwell
Welch, M. G. & Son............................Burke
Whipple, L. W. & Son.........................Malone
Whitney, C. P. ....Orleans

Will, John ....Fort Covington
Wood, J. Walter, Jr., ...Clayton
Zabriskie, Andrew C. ....Barrytown

NORTH DAKOTA.

Pope, G. Stanley ....Oacoma

OHIO.

Beatty, J. P. ....Pataskala
Betts, Henry ....Pittsfield
Cook, Howard ....Beloit
Crane, J. H. & Sons..........................Toledo
Fobes, Mrs. E. A. ....Williamsfield
Good Hold Farm ....Mentor
Greenawalt, J. S. & Son......................Beloit
Howatt, D. E. ....Cleveland
McConnell, A. B. & Son....................Wellington
Spencer, A. B. ....Goldwood
Wilson, A. J. ....Grafton

OREGON.

Honeyman, J. D. ....Portland

PENNSYLVANIA.

Arkcoll, W. W. Blake..........................Paoli
Ayer, H. S. ....Columbus
Blakeslee, O. P. ....Spartansburg
Boyer, R. A. ....Catasauqua
Butterfield, Jerome F. ....South Montrose
Byrne, Christopher ....Friendsville
Byrne, Patrick ....St. Josephs
Carrons, Robert M. ....Washington
Cass, George L. ....New Milford
Cloud, James & Son..................Kennett Square
Cornell, A. M. ....Altus
Cornell, H. S. ....Columbia Cross Roads

Davis, Edward P. ....Newtown
Deubler, James & Sons...................Springvale
Farrell, W. E. ....Corry
Friends' Asylum ....Frankford, Philadelphia
Hillview Stock Farm, Limited ....Paoli
Hopkins, Willis W. ....Aldenville
Logan, Sidney A. ....Philadelphia
McCray, C. F. & Son..........................Corry
McFadden, George H. ....Bryn Mawr
Munce, R. J. ....Washington
Oakey, John W. ....Bryn Mawr
Peck, C. L. ....Nelson
Roberts, Percival, Jr. ....Narberth
Shimer, B. Luther ....Bethlehem
Simpson, John ....Scranton
Stewart, C. E. ....Hartstown
Templeton, Robert & Son......................Ulster
Turnbull, Thomas, Jr.,....835 Western Ave., Allegheny
Valentine, John R. ....Bryn Mawr

Rhode Island.

Bowen, Edward S. ....Pawtucket
Brown, Obadiah, Estate of................Providence
Hawes, Addison S. ....Providence
Joslin, H. S. ....Mohegan
Sherman, Everett B. ....Harrisville
Sherman, Leander ....Harrisville
Smith, Benjamin F. ....North Scituate
Smith, Daniel A. ....Tarkiln
Tefft, S. Frank ....Hamilton
Vaughn, William P. ....Providence
Winsor, Nicholas S. ....Greenville

South Carolina

Clayton, B. F. & Son ....Anderson
Hinson, W. G. ....Charleston

South Dakota.

Cosgrove, Michael ....Madison

Texas.

Turner, J. C. ....Longview

Virginia.

Groome, H. C. ....Warrenton
Turnbull, Thomas, Jr. ....Casanova
Venable, A. R., Jr. ....Farmville

Vermont.

Abell, C. A. ....St. Albans
Anderson, A. J. & Son ....North Craftsbury
Buck, C. W. ....Brownsville
Butterfield, B. F. ....Derby Line
Clark, H. A. ....Hyde Park
Collins, F. O. ....St. Albans
Copeland Farm ....Middletown
Cramton, W. S. ....Rutland
Davidson, George ....South Royalton
Drew, F. A. ....South Burlington
Dunsmore, George ....Swanton
Emerson, Charles W. ....Charlotte
Fisher & May ....St. Albans Hill
Fletcher, A. M. ....Proctorsville
Forest Park Farm ....Brandon
Foss, J. Barron ....St. Albans
Hannah, Matthew ....Brownsville
Houghton, W. W. ....Lyndonville
Jackman, W. H. ....Vergennes
Joslyn, F. A. ....Northfield
LaDue, William ....Braintree, P. O. Roxbury
Lovejoy & Eddy ....Stowe
Martin, H. W. ....Bradford

Merriam, W. A. ....Glover
Nye, W. C. ....East Barre
Parker, R. & Son ....Ferrisburg
Proctor, Fletcher D. ....Proctor
Rice, George L. ....Rutland
Scott, W. F. ....Brandon
Scribner, G. S., Estate of..................Castleton
Soule, Arthur B. ....St. Albans
Spalding, F. W. ....Poultney
Stevens, C. B. ....St. Johnsbury
Stevens, William Stanford ....St. Albans
Turner, Walter D........................Moretown
Vaughan, C. A. & R. H. ....Thetford Centre
Vermont Experiment Station ....,..Burlington
Vermont Industrial School ....Vergennes
Watson, H. R. C. ....Brandon
Winslow, C. M. ....Brandon

WASHINGTON.

Clise, J. W. ....Seattle
Farrell, J. D. ....Seattle
McMillan, Gilbert ....Redmond
Stockwell, A. P. ....Aberdeen

WEST VIRGINIA.

Reymann, L. A. ....Wheeling

WISCONSIN.

Finn, James ....Whitewater
Jones, Sam ....Juneau
Lueder, R. A. ....Plymouth
Nelson, George A. ....Milltown
Seitz, Adam ....Waukesha
Tschudy, Fred ....Monroe

RESIDENCE UNKNOWN.

Angell, Edwin G.
Birnie, Charles
Bradford, A. H.
Crane, Fred
Dearborne, A. J.
Fairweather, William
Gibb, J. L.
Haskins, J. P.
Hawkes, E. B.
Hyde, J. B.
Krebs, J. DeWitt
Robinson, Isaac R.
Sadler, Edward W.
Smith, J. B.
Thorp, John C.
Thurber, C. S.
Walcott, C. W.
Wood, Lucius H.

CANADA.

Cochran, M. H. ... Compton, Que.
Clark, J. G., ... Ottawa, Ont.
Hume, Alex. & Co. ... Menie, Ont.
Hunter, Robert & Sons ... Maxville, Ont.
Irving, Thomas, ... Howick, Que.
Ness, R. R. ... Howick, Que.
Stephen, W. F. ... Huntingdon, Que.

AYRSHIRE CATTLE.

C. M. Winslow.

This breed of dairy cattle originated in the County or Ayr, Scotland, and has been known as a distinct breed for over one hundred and fifty years.

The County of Ayr is in its climate and natural resources admirably calculated to originate and foster a hardy and enduring race of cattle, being of naturally strong and productive soil, but swept by frequent coast storms of sufficient severity to try the constitution of the most hardy cattle.

The natural combination of severe climate and good food and the Scotch method of compelling the cows to roam over the heath in all weathers was calculated to develop such traits of constitution and dairy quality as would fit them for Canada and the New England states.

They were introduced into Canada by the early Scotch settlers and have ever been a favorite breed in all that northern country, and at a somewhat later period they were introduced into New England by the Massachusetts Society for the Promotion of Agriculture and by a few private individuals in New England and the Middle States, and are gradually spreading westward. The Ayrshire is found to be just the cow for the Northwest.

Origin.

Various theories have been promulgated as to the early history of the Ayrshire, but the most probable one is that they descended from the early white cattle of Great Britain. Earliest history speaks of the native wild cattle of the country as being white, with red ears and black noses, high white horns with black tips, with an animal now and then having more of the brown, black or red, very wild and the bulls fierce, but when calves are taken young, grow up to be quiet and tame. This origin seems reasonable because white is the natural color of the Ayrshire, reverting to it even under adverse conditions.

The first we hear of any effort being made to improve the native stock of the country was about 1700 and this was said to have been accomplished by selection and better care.

About 1750, we read from Aiton that Earl of Marchmont purchased from the Bishop of Durham, and carried to his seat in Berwickshire, several cows and a bull of the Teeswater or other English breed of a brown and white color.

He also writes that about 1770, bulls and cows of the Teeswater or Shorthorn breed were said to have been introduced by several proprietors, and it is from them and their crosses with the native stock that the present dairy breed has been formed.

In 1811, in "Survey of Ayrshire," Aiton writes that the Ayrshire dairy breed is "—in a great measure the native indigenous breed of the County of Ayr, improved in their size, shapes and qualities, chiefly by judicious selection, cross coupling, feeding and treatment for a long series of time and with much judgment and attention."

From about the beginning of the last century we find frequent mention of efforts for improvement in the shape of the body and especially in the shape of the udder.

Mention is made of a particular family of Ayrshires called the Swinley variety, obtained by infusion of the West Highland blood which produced cattle with a broader head, more upright horns, thicker hair and stronger constitutions.

There are traditions that an Alderney cross was made on the Dunlop strain.

Although authorities differ somewhat in regard to the detail of the methods pursued in the early improvement of the Ayrshire, still all agree that the modern Ayrshire is the result of a cross or crosses between the native cattle and some foreign breed or breeds.

IMPORTATIONS.

Ayrshire cattle were early in the past century brought over to Canada by the early Scotch settlers and on ships from Glasgow to supply the passengers with milk during the voyage and sold to the farmers on arrival at port, either at Quebec or Montreal.

They were about the same time brought to the United States by the Massachusetts Society for the Promotion of Agriculture, and let out to the farmers of the state to improve the native cattle.

Ayrshire cattle found a congenial climate in Canada and New England and were able to adapt themselves to their natural surroundings in their new homes without any great change in acclimating, and have steadily found increasing favor in any section where they have been introduced for a dairy cow, and particularly where the food supply is limited and economy of production is an object.

Recently there have been an increased number of Ayrshires imported from Scotland both into Canada and the States, and as a rule they have been selected more with a view of getting dairy animals than was done a few years ago, and selected with more attention to the teats, to have them adapted to the needs of the American dairyman.

DESCRIPTION.

The Ayrshire cow is of medium size, weighing on an average about one thousand pounds. She has a small bony head, dished face, large full eyes, large nostrils and a capacious mouth, with a broad muzzle, upright horns, a long slim neck, thin sloping shoulders, with the spine rising a little above the shoulder blades, large barrel, long and broad hips, heavy hind quarters, giving her the appearance of being wedge shape, back level to setting on of tail, except a rise at the pelvic arch, broad across the loin, barrel deep and large, with ribs well sprung to give abundant room for coarse fodder and wide through the region of the heart and lungs. Hind legs straight, thigh thin and incurving, giving room for udder, legs short, bones fine and joints firm, udder large when full and nearly level with belly; wide, long and strongly hung to the belly and thighs, teats two and one-half to three and one-half inches long, of good size, placed wide apart on the four corners of the udder, with udder level between them and not cut up; milk veins large and tortuous, entering the belly well forward toward the fore legs. Skin soft and mellow, covered with a thick growth of fine hair.

Color dark red and white, or light red (which in Scotland is called brown) and white, the relative proportion of the red to the white being variable, sometimes being nearly all red, and sometimes nearly all white. The dark color is sometimes found almost a black, and frequently a chocolate color, especially on the bulls. The white is usually a clear white unmixed with any other color, showing a line of distinct marking between the red and white.

She is quick and active in all her movements, a tough hardy cow, with a voracious appetite, eating greedily everything placed before her, good and poor, always hungry, and not at all dainty in her appetite. She is a rapid feeder, both in the stable and pasture, and when grazing takes the first that comes to her without seeming to hunt around for the sweet morsels, but acts as though she was in a hurry to fill herself, taking good grass, poor grass, or browse, whatever she can get hold of. As soon as she has her fill she goes immediately to chewing her cud, chewing rapidly, and continuously, when either lying down or walking, and will usually keep on chewing when started into a run, seeming anxious to lose no time in getting her supply of milk ready for the stable. It is perhaps this quality that makes an Ayrshire cow always look in good condition, and give a full flow of milk when other cows are thin and dropping off in their milk flow.

She is quiet and pleasant, not easily disturbed by noise or commotion in the stable, seeming not to care whether she is milked by the same person always, or by a stranger. She is bright and intelligent and is easily taught to take the same place in the stable, and readily adapts herself to a new place if desired.

She is a persistent milker, giving a large flow up to nearly time for calving, and unless care is exercised, will not readily dry off.

Milkman's Cow.

As a dairy cow she is particularly adapted to the production of milk for the milkman and table use, as her medium size, vigorous appetite and easy keeping qualities make her an economical producer, while her even, uniform production makes her a reliable supply, and the richness of her milk in total solids places her milk above suspicion from city milk inspectors.

Her milk will bear unusual transportation and handling without souring, and when poured back and forth a few times from one can to another will remix the cream and milk, which will not again readily separate, giving it a uniformity in quality until the last is sold or used. It has a good body and is rich looking, never looking blue.

The milk itself being evenly balanced with casein and butter fat is a complete food, easily digested, nutritious and is particularly adapted to children and invalids. Stomachs that are weak and unable to digest other milk find no trouble with Ayrshire cow's milk.

MILK AND BUTTER RECORD.

The capability of the Ayrshire cow as shown by official records, made under the supervision of the Ayrshire Breeders' Association, and States' Experiment Stations. For Milk and Butter.

Bulls admitted to the Advanced Register.

Nonpareil 4535, Adv. R. 1,	with 2 tested daughters	
Reynard 6038, Adv. R. 2,	" 6 "	"
Linwood 4213, Adv. R. 3	" 2 "	"
Duke of Ayer 6180, Adv. R. 4,	" 6 "	"
Glencairn of Ridgeside 6248, Adv. R. 5,	" 2 "	"
Nox'emall 7312, Adv. R. 6,	" 9 "	"
Moonstone of Drumsuie 8228, Adv. R. 7,	" 11 "	"
Carbello 6634, Adv. R. 8,	" 3 "	"
Geo. of Rosemont 7670, Adv. R. 9,	" 7 "	"
Rothage 6044, Adv. R. 10,	" 2 "	"
Oshawa of Highland 7225, Adv. R. 11,	" 4 "	"
Twister 5651, Adv. R. 12,	" 3 "	"
Major Ayer 5533, Adv. R. 13,	" 7 "	"
Colonel Ayer 7168, Adv. R. 14,	" 11 "	"
Prince of Barclay 6711, Adv. R. 15,	" 5 "	"
Sebastian 6269, Adv. R. 16,	" 6 "	"
Beloit Ayer 6775, Adv. R. 17,	" 2 "	"

Calmar 4692, Adv. R. 18, with 2 tested daughters
Earl's Choice of Spring Hill 8289, Adv. R. 19, " 5 " "
Jo Smith 7398, Adv. R. 20, " 2 " "
John Doland 8772, Adv. R. 21, " 3 " "
Yucca Lad 6878, Adv. R. " 3 " "
Harcourt of B 7272, Adv. R. " 2 " "

Cows Admitted to Advanced Registry.

Seven Day Records.

	Lbs. of milk	butter fat	age
Lucretia B of Riverside 21597, Adv. R. 32	227	8.60	2¾ yrs.
Beauty B 22741, Adv. R. 33	248	9.43	2¾
Fleet Melrose 18980, Adv. R.	303	11.48	3¾
Roseleaf Douglas 13449, Adv. R. 34	357	12.86	Mature
Annie Bert 9670, Adv. R. 35	356	12.43	"
Ouija 11882, Adv. R. 36	378	13.20	"
Kirkland Sparrow 20951, Adv. R. 37	382	15.23	"

Year's Records.

Two year old Heifers, arranged according to age.

Blossom of Seattle 20615, Adv. R. 38	6,027	256.30	1¾ yrs.
May Rose of Radnor 18695, Adv. R. 39	6,898	288.00	1¾
Florine Corslet 17512, Adv. R. 40	5,504	212.60	1¾
Bonnie 2d of Radnor 19754, Adv. R. 41	8,184	345.40	1¾
Muriel Girl 18264, Adv. R. 42	5,914	244.30	1¾
Lilac of Radnor 18690, Adv. R. 43	7,778	300.90	1¾
Miss Kilbowie 17505, Adv. R. 31	6,751	236.00	2¼

Rose Ascott 15035, Adv. R. 30.. 5,621 207.43 2¼
Bessie of Rosemont 17904, Adv.
    R. 44 .... 8,835 377.11 2¼
Rose Crashaw 17507, Adv. R. 45 5,995 230.57 2¼
Ruth 2d of Barclay 19753, Adv.
    R. 46 .... 6,338 291.43 2¼
Sibyl Corslet 18256, Adv. R. 47 7,170 271.71 2¼
Lady Wonder 4th 18043, Adv. R. 48 5,606 210.00 2¼
Crimson Maid 22,763, Adv. R. ... 7,090 283.83 2¼
Douglas Daisy Queen 24883,
    Adv. R. 7,886 280.70 2¼
Ruby Douglas 16672, Adv. R. 29 6,321 252.00 2¼
Muriel Fox 15036, Adv. R. 28.. 6,685 264.00 2¼
Rose Clockston 15026, Adv. R. 27 6,135 209.14 2¼
Myrtle K 19615, Adv. R. 49.... 7,497 251.14 2¼
Isabella of Sand Hill 20366, Adv.
    R. 50 .... 7,887 319.71 2¼
Rose Brodick 15029, Adv. R. 26 7,117 266.60 2¼
Rose Aileen 18255, Adv. R. 51.. 6,256 212.60 2¼
Bessie of Sand Hill 20365, Adv.
    R. 52 .... 7,307 322.30 2½
Rose Foxglove 15038, Adv. R. 25 6,128 282.97 2½
Rose Claymore 17511, Adv. R. 53 6,542 269.14 2½
Pearl Douglas 17453, Adv. R. 54 6,598 271.71 2½
Buttercup of Rosemont 17900,
    Adv. R. 55 .... 7,584 305.14 2½
Bell Ayer 20180, Adv. R. 56.... 7,111 309.51 2½
Francis of Barclay 18687, Adv.
    R. 57 .... 8,047 345.43 2½
Mina 3d of Radnor 20880, Adv R. 8,349 382.32 2½
Queen 2d of Barclay 19748,
    Adv. R. 58 .... 9,486 364.28 2½
Grace of Sand Hill 22472, Adv.
    R. 59 .... 8,698 338.57 2½
Myrtle Kilbowie 18262, Adv. R. 60 7,199 281.16 2½

Crimson Rambler 21109,Adv.R.61	7,988	278.57 2½
Mayflower of Seattle, 20614, Adv. R. 62	6,496	250.30 2½
Mattie of Sand Hill 23323, Adv.R.	8,212	349.56 2½
Lulu Avondale 15053, Adv. R. 24	6,122	258.30 2½
Mosshawk of Barclay 19746, Adv. R. 63	6,086	256.30 2½
Felicia of Woodview 17431, Adv. R. 64	7,048	279.43 2½
Spotted Lola 20876, Adv. R.	6,924	289.22 2½
Dora of Sand Hill 23322, Adv R.	7,622	312.89 2¾
Floe 16700, Adv. R. 23	8,201	345.43 2¾
Queen of Seattle 20613,Adv.R. 65	7,896	318.00 2¾
Ruby Russell 3d 19936,Adv. R. 66	6,760	244.30 2¾
Dorinda 20902, Adv. R.	7,855	302.77 2¾
Beulah M 20368, Adv. R.	8,516	357.53 2¾
Dolly Fryer 2d 17094, Adv. R. 67	6,485	256.30 2¾
Petrina of Woodview 17430, Adv. R. 22	7,766	346.00 2¾
Kitty K 3d 21246, Adv. R. 68	7,364	329.14 2¾
Ruth of Sand Hill 22473, Adv. R. 69	7,699	309.43 2¾
Rose Dolman 13688, Adv. R. 21	7,409	268.30 2¾
Clotilde of Rosemont 17893, Adv. R. 70	8,548	322.30 2¾
Baby Douglas 21849, Adv. R. 71	9,652	377.30 2¾
Rena Ayer 21247, Adv. R. 72	6,929	306.00 2¾
Snowflight of Radnor 18696, Adv. R. 73	8,505	349.30 2¾
Oshawa Lass of Highland 2d 20179 Adv. R. 74	6,281	263.14 2¾
Lizzie Muriel 15364, Adv. R. 20	7,583	287.14 2¾
White Lola 19756, Adv. R. 75	8,282	295.71 2¾
Eunie of Highland 21694,Adv. R.	8,093	354.79 2¾
Megsy Tipperlin 17432,Adv.R. 76	7,117	270.00 2¾

Princess of Highland 2d 21250
    Adv. R. 77 7,795   321.43 2¾
Oshawa Lady 2d 18249,Adv.R. 78 7,074   305.14 2¾
Colonel's Daughter 21254,Adv. R. 6,565   255.59 2¾
Douglas Cordelia21852,Adv.R. 79 8,649   298.30 2¾
Katy Did 15242, Adv. R. 19.... 6,760   280.30 2¾
Rose Eaton 20511, Adv. R. 80... 7,783   277.71 2¾
Starlass 20182, Adv. R......... 7,095   268.08 2¾
Letta Lind of Radnor 17892,
    Adv R. 81 8,602   372.85 2¾
Felicia of Highland21754,Adv.R. 8,767   344.27 2¾

THREE YEAR OLD FORM

Lady Rotha 18821, Adv. R. 82.. 8,101   315.42 3¼
Eugenie Douglas, 17452, Adv.
    R. 83 9,587   379.71 3¼
Belle's Cherry 15263, Adv. R. 18 8,871   360.85 3¼
White Bess of Radnor 18689,
    Adv. R. 84 6,594   262.30 3¼
Orinda 18824, Adv. R. 85...... 7,375   280.30 3¼
Ithan 3d 21253, Adv. R......... 8,360   347.24 3¼
Kaziah 2d 20181, Adv. R....... 7,961   294.87 3¼
Stilletto 16701, Adv. R. 86...... 6,707   263.14 3¼
Ponemah 2d 17614, Adv. R. 87.. 7,330   258.00 3¼
Neidpath Lassie 2d 18248, Adv.
    R. 88 6,746   258.00 3¼
Midget of Sand Hill 19487, Adv.
    R. 89 9,824   361.71 3¼
Cora T 3d 19489, Adv. R. 90... 7,312   301.71 3¼
Lady Bell 4th 17256, Adv. R. 91 8,516   374.57 3¼
Kitty K 2d 18247, Adv. R. 92.... 8,255   352.30 3¼
Cherry of Barclay 19752, Adv. R. 8,599   371.18 3¼
Curfew Bell 21255, Adv. R.....11,181   502.99 3½
Oshawa's Prudence 21248,Adv.R. 7,668   313.45 3½
Oshawa's Pride, 21249 Adv. R.. 7,681   319.25 3½
Chatauqua Fairy19350,Adv.R. 93 7,524   305.14 3½

Angeline Sebastian 18681, Adv.
R. 94 .... 8176 340.30 3½
Rotha of Ridgeside 17360, Adv.
R. 95 .... 7,324 330.86 3½
Doris Y 16351, Adv. R. 96 .... 7,807 313.00 3½
Yucca Douglas 19504, Adv. R... 9,049 317.77 3½
Daisy Jewess 3d 20324, Adv. R. 97 9,665 356.57 3½
Lary Burton 19507, Adv. R..... 9,580 326.95 3½
Madonna Lass 3d 21850, Adv. R. 10,467 429.86 3½
Soncy of Barclay 19755, Adv. R. 7,680 320.67 3½
Drummond's Gem 22652, Adv. R10,746 356.30 3½
Sweet Josie 19833, Adv. R. 98..10,103 404.57 3¾
Broomhill Dairymaid 22218, Adv.
R. 99 .... 8,326 318.85 3¾
Jennie of Sand Hill 19490, Adv.
R. 10,160 437.17 3¾
Nellie of Highland 17255, Adv.
R. 100 ..,.............. 8,374 333.43 3¾
Barleith Snowdrop 22219, Adv.
R. 101 .... 8,158 299.13 3¾
Babe's Duchess 22213, Adv.R.102 9,559 387.31 3¾

FOUR YEAR OLD FORM.

Abbie Sebastian 20531, Adv. R.
103 ....10,449 366.00 4¼
Holehouse White Bess 22221
Adv. R. 7,981 347.87 4¼
Auchenbrain Princess 7th 21622,
Adv. R. 104 .... 8,485 355.71 4¼
Finlayston Maggie 3d 19217,
Adv. R. 105 ....10,759 439.71 4¼
Broomhill Minnie 10th 21627,
Adv. R. 106 .... 9,409 380.57 4¼

Agnese Sebastian 18679, Adv.
R. 107 .... 9,364 366.00 4¼
Ithan 2d 17254, Adv. R. 108.... 8,174 368.57 4¼
Pauline Sebastian 18678, Adv.
R. 109 ....10,745 361.71 4¼
Madonna Lass 2d 17473, Adv. R.
110 ....10,020 384.85 4¼
Bessie of Rosemont 17904, Adv.
R. 44 ....14,102 578.57 4¼
Garclaugh Mayflower 20956, Adv.
R. 9,814 341.65 4¼
Douglas Empress 21851, Adv. R. 9,165 382.87 4½
October Lass 20323, Adv. R. 111 10,078 397.43 4½
Polly Sebastian 20,529, Adv. R... 8,189 315.33 4½
Dollie Kilbowie 16779, Adv. R... 9,039 311.14 4½
Lulu Avondale 15033, Adv. R.
24 (2d entry) 8,326 336.00 4½
Miss Betty of Spring Hill 17997,
Adv. R. 113 .... 8,334 304.28 4½
Esthon Douglas 19153, Adv. R...10,527 373.52 4¾
Daisy Jewess 2d 17472, Adv.
R. 114 .... 8,574 336.86 4¾
Cora T 2d 18400, Adv. R.......12,230 510.67 4¾
Beckey of Holehouse 17015, Adv.
R. 115 ....10,507 396.85 4¾
Lizzie of Barclay 17024, Adv. R.
116 .... 8,868 408.00 4¾

MATURE COWS.

	Lbs. of milk	butter fat
Rena Myrtle 9530, Adv. R. 1	12,172	468.00
Atalanta 10777, Adv. R. 2.............	9,740	367.71

Nancy B. 9581, Adv. R. 3	8,782	356.56
Miss Ollie 12039, Adv. R. 4	9924	440.60
Durwood 12680, Adv. R. 5	10,701	433.71
Cad's Beauty 13606, Adv. R. 6	8,702	382.30
Himona 13032, Adv. R. 7	8,765	376.30
Acelista 12094, Adv. R. 8	9,906	361.71
Belle Nixon 14705, Adv. R. 9	9,383	360.85
Roanette 11476, Adv. R. 10	8,638	331.71
Xoa 11469, Adv. R. 11	9,090	331.71
Lukolela 12357, Adv. R. 12	9,299	329.14
Yucca 11470, Adv. R. 13	8,502	323.14
Inez Douglas 14554, Adv. R. 14	9,089	426.00
Durline 13473, Adv. R. 15	9,317	345.43
Mysie of Barcheskie 14952, Adv. R. 16	9,228	336.85
Iola Lorne 12773, Adv. R. 17	8,806	335.14
Rose Pender 18645, Adv. R. 117	9,913	356.55
Ada Rome 17461, Adv. R. 118	9,835	351.43
Ruby Russell 15564, Adv. R.	8,643	327.43
Molly Fryer 16051, Adv. R. 120	9,741	388.30
Lady Sam 16286, Adv. R. 121	9,530	349.00
Acelista 12094, Adv. R. 8 (2d entry)	11,856	419.14
Iola Lorne 12773, Adv. R. 17 (2d entry)	9,675	337.71
Alcyone of the Plain 13318, Adv. R. 122	8,646	323.14
Polly of Mauchlin 17861, Adv. R. 123	9,321	364.30
Lillian Drummond 4th 16189, Adv. R. 124	9,239	369.43
Hetty Ayer 14030, Adv. R. 125	9,858	474.85
Bertha M 15262, Adv. R. 126	8,754	366.00
Etta Poultney 18148, Adv. R. 127	11,475	419.14
Avis 16152, Adv. R. 128	9,424	324.00
Crimsonia 2d 13715, Adv. R. 129	9,645	358.57
Mary A. M. 2d 16466, Adv. R. 130	9,782	343.00
Clarissa Lorain 16538, Adv. R. 131	9,051	329.14
Miss Mabel D 15874, Adv. R. 132	9,693	349.71
Polly Puss 16296, Adv. R. 133	12,632	500.57
Fern Ayer 16289, Adv. R. 134	13,601	519.64

Rena Webb 12479, Adv. R. 135	9,366	364.28
Oshawa Lady 16020, Adv. R. 136	9,695	346.28
Lady Bell 2d 16536, Adv. R. 137	8,628	356.57
Oshawa Lass of Highland 16534, Adv. R. 138	8,561	324.00
Kitty K 12933, Adv. R. 139	11,115	439.00
Keepsake 15913, Adv. R. 140	10,868	439.71
Rena Ross 14539, Adv. R. 141	10,065	439.00
Ithan 14538, Adv. R. 142	9,975	396.85
Flora 4th of Bonshaw 15578, Adv. R. 143	9,874	373.00
Frisky of Bonshaw 17018, Adv. R. 144	8,767	373.71
Maggie of Radnor 17013, Adv. R. 145	9,468	406.33
Denty 9th of Auchenbrain 15577, Adv. R. 146	11,757	452.57
Daisy of Rosemont 17011, Adv. R. 147	9,164	340.30
Queen of Barclay 15096, Adv. R. 148	9,172	342.00
Lily 4th of Fairfield Mains 15579, Adv. R. 149	9,054	343.00
Brown Eyes of Knockdon 19216, Adv. R. 150	8,724	354.00
Lady Browning 15105, Adv. R. 151	8,531	321.43
Frisky of Bonshawl 7018, Adv. R. 144 (2d entry)	9,645	396.86
Queen of Barclay 15096, Adv. R. 148 (2d entry)	11,258	414.86
Clockston Bella 2d 21628, Adv. R. 152	8,509	353.14
Miss Olga 13984, Adv. R. 153	10,200	386.57
Biona 12351, Adv. R. 154	10,012	337 71
Finlayston Cherry 6th 21427, Adv. R. 155	8,903	336.00
May Mitchel 20922, Adv. R. 156	9,419	356.00
Rena Ross 14539, Adv. R. 141	15,072	643.71
Hillside of Toledo 16022, Adv. R. 157	9,179	330.00
Addington Queen 18949, Adv. R. 158	9,348	390.85
Lady Bruce 21704, Adv. R.	8,648	326.47
Cambric 18664, Adv R	9,745	341.29

Agnese Sebastian 18679, Adv. R. 107 (2d
 entry) 8,660 393.64
Megsy Tipperlin 17432, Adv. R. 76 (2d
 entry) 10,701 399.32
Princess Beatrice 2d 16533, Adv. R....... 9,445 372.92
Woodside May 21703, Adv. R........... 9,042 377.31
Nino Douglas 15959, Adv. R. 11,115 431.60
Dewdrop of Spring Hill 20950, Adv. R...10,554 391.73
Primrose 20970, Adv. R.................10410 386.04
Lady Belle of Spring Hill 18000, Adv. R. 9,466 362.66
Oshawa Lady 2d 18249, Adv. R. 78 (2d
 entry) 11,254 430.73
Keepsake 15913, Adv. R. 140 (2d entry) 12,365 469.32
Brown Eyes of Knockdon 19216, Adv. R.
 150 (2d entry) 11,448 437.61
Cora T. 13772, Adv. R................12,392 448.99
Auchenbrain White Beauty 2d 21687,.
 Adv. R. 10,494 438.32
Princess Lily 21,702, Adv. R............11,771 438.03
Bell Ayer 20180, Adv. R. 56 ....11,934 492.91
Bessie Clyde 17287, Adv. R.............12,044 459.21
Auchenbrack Sweet Pea 2d 21625, Adv.
 R. 13,097 532.87

CUTTING FROM A SCOTCH PAPER.

TESTS OF MILK COWS IN AYRSHIRE.

Mr. John Speir, Newton farm, Newton, Glasgow, gave a lecture on "Tests of Milk Cows in Ayrshire" in the Town Hall, Cumnock, yesterday afternoon. There was a good attendance of farmers, who showed much interest in, and appreciation of, the lecture.

Mr. Speir said:—For three years the Highland and Agricultural Society have been doing good work in the endeavor to encourage the owners of herds of Ayrshires to systematically record the produce, in regard to both quality and quantity, for the whole of their milking period. During the season 1905 they had 815 cows under continuous supervision in the districts of Cumnock and Fenwick. The result of this testing is the most conclusive yet carried out by the Society of the great value of selecting cows and bulls for breeding purposes only from mothers which have proved themselves good milkers. From the records of the past year, it is easily seen that among the Ayrshires there are many better milking animals than even the most ardent advocates of the merits of the breed ever anticipated. What, however, much reduces the value of the breed for dairy purposes is the great irregularity in milk yield of many families. It has been long proved beyond a doubt that the ability to yield a large quantity of milk is an inherited qualification, just as much as any of the other items of the nature of each individual animal. Food, provided it is in moderate quantity, has little to do with it, as is shown by the records of the Cumnock district, where the farms are all very much of one class. There, out of 372 cows under test, the ten heaviest milking cows were all in one herd. At one time it was supposed that a rich quality of milk must necessarily follow the use of rich food. Repeated experiments during the past 15 years have demonstrated that quality in milk can be little altered by feeding. These tests clearly indicate the same, as cows of the same age, going on the same pasture, and in other respects treated alike, yielded milk from 50 to 60 per cent. richer in fat than others of a different family. The same applies to the animals when in the house, where the food was more under control than in the field. Under these circumstances many of the heaviest fed stocks gave not only the poorest milk, but least of it. In the Fenwick district, where the records were carried on during the whole year, out of 443 cows, there were 9, or 2 per cent., which yielded over £30 in milk; 37, or 8 per cent., which yielded over £25 in milk; and 137, or 31 per cent., which yielded over £20 in milk. Against that there was a considerable number which yielded only from £8 to £11 of milk in the year, the milk being valued at 5d. per gallon for milk of 3 per cent. of fat. In the Cumnock district the supervision of the herds was only continued for thirty-four weeks, and while the milk yield for the period was just as good as in the Fenwick district, it does not total up to such a large figure. In that period one cow yielded on the grass milk of a total value of £26, 7s. 11d., and other ten of upwards of £20. Among the heifers tested at Cumnock were some particularly good ones, about a dozen having yielded milk of a value of from £14 to £16, 10s. in thirty-four weeks, with, in many cases, a very large quantity not only before testing began, but after it ceased. When these results are compared with the milk yield of others, which only had a value of from £6, 10s. to £9, 10s. during the season, the value of the method suggested for the selection of cows for breeding purposes is at once seen. A very instructive wall diagram was exhibited showing the yield and value of 10 per cent. of the best and worst of the cows of each herd. Out of 372 cows, there were 35 which yielded milk of an average value of £17, 4s. 2d., while there was an equal number, the milk of which was only of the value of £10, 19s. 2d. The difference is £6, 5s. between these two lots of cows in thirty-four weeks, which for the whole milking period might probably be £8 or £10, as when the testing stopped many of the best milking cows were giving a considerable quantity of milk, or, as previously stated, had already done so, while the poor ones were mostly dry. It was also worthy of note that the best and heaviest milking cows usually gave the richest milk. In an odd instance or two this did not occur, but as a rule it did so. Wall diagrams were exhibited showing most of the principal details, each farm being indicated by a letter. As showing the popularity of this work, on which the Highland and Agricultural Society is spending about £200 each year, it may be stated that there are a great many inquiries from buyers of bulls, who wish them out of cows which can be certified to have given a certain quantity and quality of milk. Each farm has a book with the details of each cow, so that the owners of the good ones can easily show a buyer their record, while those having cows giving a low yield are not penalized by their names being published. During the present year there are five milk record societies at work under the Highland and Agricultural Society's scheme, which for this season will control about 2,000 cows.

FORMALDEHYDE AS A REMEDY FOR CALF SCOURS.

I have used this remedy with satisfactory results, and believe it to be the best remedy I have ever tried.

C. M. WINSLOW.

The Maryland station grappled with the question of calf scours, a disease that takes off millions of dollars worth of live stock every year. A late bulletin announces that they have found that 1 part of formalin to 4,000 parts of milk will almost invariably destroy the organisms in the bowels of the calf that produces the disease. Eleven out of twelve calves responded favorably to this treatment. The formula is to dissolve ½ ounce of formalin in 15½ ounces of water and add a teaspoonful of this liquid to each pound of milk fed the calf.

MAINE AND CONNECTICUT

Maple Grove Farm
... Ayrshires ...

Our herd is now headed by Imp. Barcheskie's Copestone (100 34) Grand Champion at Brockton, 1907. We are breeding for utility, good size and dairy quality. Cow from our herd won the single cow dairy test for pounds of milk and fat in one day at the Maine State Fair 1908. All breeds competing. Herd Tuberculin tested. Stock of all ages for sale.

JOHN A. & ROWLAND NESS, Auburn, Me.

— Homehill Farm —

Ayrshires for Business

Bulls from Advanced Registry Stock for sale.

Our two-year-old heifers are earning from $125.00 to $175.00 per year. Our mature cows from $160.00 to $200.00 per year. Our milk is sold in the wholesale market, averaging 3 1-2 cents per quart for the year.

Our stock bull is Mauchlin of Homehill, No. 10365, grandson of Sir Thomas of Auchenbrain No. "2760", Scotch Record.

Write for Official records and prices.

Henry Dorrance, Prop.

Plainfield, - - **Conn.**

VERMONT

1873--1909

C. M. Winslow,

BRANDON, - - VERMONT

N. Y. C. R. R., Rutland Division

Farm Situated Near
The Railroad Station

Herd Free From Tuberculosis

This herd was established in 1873 by purchase of a few heifers from cows that had strong constitutions, were typical animals of the breed, had shapely udders and long teats.

These families have been bred on the farm since then, with care in selecting bulls, to perpetuate the desirable qualities found in the Ayrshire cow.

As the herd stands today, they are dark red and white, white and dark red, small upright horns, shapely bodies, of good size, typical udders with long teats and have a quiet disposition.

They have been bred from the start for a profitable dairy herd that should be good feeders and good milkers.

YOUNG STOCK FOR SALE

VERMONT

HILLCROFT STOCK FARM
Brownsville, Vermont

...60 Head Registered Ayrshires...

FRISKEY 9221

This herd has won over 900 premiums in five years, including champion and grand champion prizes, also milk and butter tests over all breeds at the Springfield, Vt., and Woodstock, Vt., fairs.

Champion Ayrshire Cow for two years at Vermont State Fair.

Visitors Welcomed

Young stock for sale from 60 lb. cows, that have goods udders, good teats, and popular color.

ISALEIGH DON 10710

Matthew Hannah.

VERMONT AND NEW YORK

BROOKLAWN HERD

F. W. SPALDING,

Poultney, - Vermont

Herd Established by the Late
L. C. Spalding in 1869

STOCK FOR SALE

L. W. WHIPPLE & SONS
Malone, N. Y.

Herd of Ayrshires are all direct from late Scotch Importations.

We have spared neither pains nor expense to get the best. We shall have a limited number of calves of both sexes for sale at very reasonable prices considering the breeding.

Let us select for you and we are sure we can please you with quality and surprise you in price

E. TEN EYCK LANSING

OAKHURST STOCK FARM, LITTLE FALLS, N. Y.

Mem. Ayrshire Breeders' Association, Mem. D. H. S. B. A. of Am.

Herd headed by Acelista's Lad 10285 out of Acelista 12094. Champion 5 yr. test produced 52,000 lbs. of milk, yielding 2137 lbs. of butter

BULL CALVES FOR SALE AT REASONABLE PRICES.

NEW YORK

Sunnyside Stock Farm

HOME OF THE AYRSHIRES

Herd bull Jean's Canuck 10011, his dam Jean Armour 18165, well known to Ayrshire Breeders. Cows of the Ayer family close bred to Major Ayer, the sire of Rena Ross and Keepsake ::

A few young things for sale.

C. W. Lewis & Son,

Alfred Station, - - **N. Y.**

Maple Row Stock Farm
... AYRSHIRES ...

We have 75 head of Ayrshires of all ages and keep none that will not fill the bill in the dairy, as they must be milkers or leave the farm

Stock of all ages for sale at reasonable prices.

Correspondence Solicited.

F. H. Cookingham,

Cherry Creek, - - **N. Y.**

… NEW YORK

Clover Home Farm

Gouverneur, New York

Registered Ayrshires

The home of over sixty head of Ayrshires that have been bred and fed for profitable dairy results

We breed for size as well as all dairy qualities and haven't a short-teated cow in the herd.

We maintain at the head of the herd two bulls of the best breeding we can obtain. Bulls now heading are:-

JACK MACDONALD No. 10259 a grandson of the imported DENTY 9th of Auchenbrain, whose official record is 11,757 lbs. milk 528 lbs. butter

PRE-EMINENT No. 11637 a great-grandson of the famous butter cow-MISS OLLIE. This youngster is of our own breeding and is the best type of a truly dairy bull we ever saw. He will be bred to a few cows and to some young heifers not related and we can promise good results

HERD TUBERCULIN TESTED

George E. Pike,
Gouverneur, New York

Long Distance Bell Telephone.
Reference, Bank of Gouverneur.

M. J. Karr　　　　　　S. S. Karr　　　　　　I. D. Karr

Sand Hill Stock Farm
S. S. KARR & SONS
Allegany Co., Almond, N. Y.

The following is a list of our cows that have made Advanced Registry records during the past two years:-

Name	Age	Milk	Butter
Isabella of Sand Hill	2 yr.	7887 lbs.	373 lbs.
Bessie of Sand Hill	2 "	7307 "	376 "
Grace of Sand Hill	2 "	8698 "	395 "
Midget of Sand Hill	3 "	9824 "	422 "
Cora T 3rd	3 "	7312 "	352 "
Miss Betty of Spring Hill	4 "	8334 "	355 "
Hetty Ayr	10 "	9858 "	554 "
Bertha M	9 "	8754 "	427 "
Ruth of Sand Hill	2 "	7699 "	361 "
Beulah M	2 "	8516 "	417 "
Jennie of Sand Hill	3 "	10160 "	510 "
Cora T 2nd	4 "	12230 "	596 "
Lady Bell of Spring Hill	5 "	9466 "	423 "
Cora T	10 "	12392 "	524 "
Bessie Clyde	6 "	12044 "	536 "

Jennie of Sand Hill has the champion 3 yr. old record for butter.

The five two-yr.-old heifers we have tested averaged 8021 lbs. milk and 384 lbs. butter.

Of the seven cows in the Advanced Registry with record above 12000 lbs. milk, three of them were tested and are still owned at Sand Hill Stock Farm.

Earl's Choice of Spring Hill 8289, A. R. No. 19, is still at the head of our herd. His stock is as good as ever. He has six daughters in the Advanced Registry..

Our junior bull, White Prince of Sand Hill, is a very fine young fellow. He has lots of quality and excellent breeding. His dam, Beulah M., is an A. R. cow. He is a great-grand-son of Major Ayr. He is a grand-son of Lessnessock King of Beauty.

We have a few cows for sale, also three or four fine bull calves.

We are located on the Erie R. R. about one mile from Almond station, and five miles from city of Hornell.

Herd Tuberculin Tested.　　　　　　**Visitors Welcome.**

NEW YORK

... AYRSHIRES ...
HERD ESTABLISHED IN 1871

All ages, both sexes, bred for practical dairy purposes, size, constitution, disposition, style, length of teat and deep and persistent milkers.

SINGLE ANIMALS OR CAR LOTS

F. D. & E. STOWELL, Successors to L. D. Stowell

Black Creek, **Allegany Co.,** **N. Y.**

Elm Valley Stock Farm
HERD ESTABLISHED IN 1882

Bred for quality and quantity, good size and long teats. Bulls used in herd, Major Clyde, Major Ayr, Clarence Star and Statesman of Springhill. 40 head young stock for sale.

W. G. TUCKER, Elm Valley, N. Y.

PENNSYLVANIA

BARCLAY FARM AYRSHIRES

This herd is headed by Moonstone of Drumsuie, No. 8228, A. R. bull No. 7, sire of eleven in the list. Sire of Bessie of Rosemont, No. 17904, champion four-year-old, official record 14,102 lbs. Milk and 675 lbs. Butter in one year. Sire of Mina 3rd of Radnor, No. 20880, who has just placed the two-year-old butter record up to 446 lbs. Sire of Denty 10th, No. 20872, who at three years of age won the milk and butter test over all breeds at the Arizona State Fair in 1908. Sire of Maggie 2nd of Radnor, the two-year-old heifer that won the butter fat test in the heifer class, over all breeds, at Illinois State Fair, 1907.

Moonstone is now being assisted by the young champion Howies Majestic, No. 10000, a fine specimen of the breed and from a strong milking cow.

We have won the first prize three years in succession on a herd of five cows in the Home Dairy tests. We also won at the same time one first, two second and two third prizes on single cows. While these tests were being conducted we placed 31 cows and heifers and three of the stock bulls that have been used in this herd in advanced registry. The official record follows:

Name	No.	Age	Lbs. Milk	Lbs. Butter
Bessie of Rosemont	17904	4 yrs.	14,102	675
Auchenbrack Sweet Pea	21625	5 "	13,112	609
Denty 9th of Auchenbrain	15577	9 "	11,757	528
Friskey of Bonshaw	17018	5 "	8,775	440
" "	17018	6 "	9,553	457
" "	17018	7 "	11,419	616
Brown eyes of Knockdon	19216	7 "	8,724	413
" " "	19216	8 "	11,328	505
Queen of Barclay	15096	7 "	9,172	399
" " "	15096	8 "	9,094	418
" " "	15096	9 "	11,162	483
Finlayston Maggie 3rd	19217	4 "	10,750	513
Beckley of Holehouse	17015	4 "	10,507	463
Flora 4th of Bonshaw	15579	9 "	9,874	435
Maggie of Radnor	17013	4 "	9,468	474
Broomhill Nunnie 10th	21627	4 "	9,409	444
Daisy of Rosemont	17011	7 "	9,164	397
Lizzie of Barclay	17024	4 "	8,864	474
" " "	17024	5 "	8,858	431
" " "	17024	6 "	9,107	451
Lily 4th of Fairfield Mains	15570	9 "	9,054	400
Finlayston Cherry 6th	21427	6 "	8,903	392
Clockston Bella 2nd	21628	6 "	8,507	412
Lady Browning	15105	8 "	8,531	375
Auchenbrain Princess 7th	21425	4 "	8,485	415
Holehouse White Bess	22221	4 "	8,104	399
Broomhill Dairy Maid	22218	3 "	8,326	372
Barleith of Snowdrop	22219	3 "	8,185	349
White Bess of Radnor	18689	3 "	6,594	306
Bessie of Rosemont	17904	2 "	8,835	435
Letta Lind of Radnor	17892	2 "	8,602	435
Mina 3rd of Radnor	20880	2 "	8,349	445
Clotilde of Rosemont	17893	2 "	8,048	376
Frances of Barclay	18687	2 "	8,047	403
Lilac of Radnor	18690	2 "	7,779	351
Buttercup of Rosemont	17900	2 "	7,585	356
Spotted Lola	20876	2 "	6,924	393
May Rose of Radnor	18695	2 "	6,898	335

Twenty of the above cows are still in this herd. We have added a new importation this winter of our own selection, mostly from families that we have tried out. The entire increase this year will be for sale and we have forty head of young stock, both sexes, for sale now.

HERD TUBERCULIN TESTED EVERY YEAR

The nearest station to Barclay Farm is Rosemont, Pa., eleven miles from Philadelphia, on Main Line Penn. R. R.

Telephone 284-D-Bryn Mawr. **J. W. OAKEY, Mgr., Bryn Mawr, Pa.**

PENNSYLVANIA

HIGHLAND FARM

JOHN R. VALENTINE, Proprietor

Home of Rena Ross and Polly Puss, the Champion Ayrshire cows for milk and butter of the world. Keepsake, Home Dairy Test winner of 1906. Herd headed by two great sires. Colonel Ayer, 7168 (Advanced Registry) half brother to Rena Ross, Polly Puss, Keepsake and Fern Ayer. Sire of twelve daughters in the list. Imported Finlayston 8882, dam Finlayston Maggie 3rd 19217, ex-champion four-year-old cow of the world for milk and butter and sold at auction for $600.00.

Herd tuberculin test every nine months by the Pennsylvania Live Stock Sanitary Board. All young animals vaccinated to immune them from tuberculosis.

The following are the official Records of the herd since 1905.

Name	No.	Lbs. Milk	Lbs. Butter
Rena Ross	14539	15,072	751
" " Home Dairy Test Record		12,937	652
" " Previous Record		10,065	512
Polly Puss	16296	12,632	584
Kitty K.,	12933	11,115	512
Keepsake	15913	10,868	513
" Now on test will beat previous record			
Megsey Tipperlin	17432	10,700	466
Ithan	14538	9,975	463
Fern Ayer	16289	9,847	444
" on test again, will make over		13,000	600
Oshawa Lady	16020	9,695	404
Princess Beatrice 2nd	16533	9,445	435
Rena Webb	12479	9,316	425
Lady Bell 2nd	16535	8,628	416
Oshawa Lass of Highland	16534	8,561	378
Four Years Old.			
Ithan 2d	17254	8,174	430
Three Years Old.			
Lady Bell 4th	17256	8,516	437
Nellie of Highland	17255	8,374	389
Ithan 3rd	21253	8,360	405
Kitty K., 2nd	18247	8,255	411
Kaziah 2nd	20181	7,961	344
Oshawa Prudence	21248	7,768	366
Oshawa Pride	21249	7,681	371
Colonels Lady Bell	21251	6,910	358
Two Years Old.			
Princess of Highland 2nd	21250	7,795	376
Kitty K., 3rd	21246	7,364	384
Bell Ayer	20180	7,111	361
" Now on test, will make over		11,000	500
Starlass	20182	7,095	313
Oshawa Lady 2nd	18249	7,074	356
" " Now on test will make over		11,000	475
" " In one day she has made		66.6	
" " In one month she has made		1,922	81
" " As a four-year-old			
Felicia of Woodview	17431	7,049	326
Rena Ayer	21247	6,929	357
" On test, will beat		10,000	500
Neidpath Lassie 2nd	18248	6,742	301
Colonel's Daughter	21254	6,565	298
Oshawa Lass of Highland 2nd	20179	6,281	307

Farm located nine miles west of Philadelphia, on the Main Line of the Penn. R. R.

PHILIP C. PALMER, V. S., Mgr.

Highland Farm - - Bryn Mawr, Pa.

PENNSYLVANIA

Penshurst Farm
Narberth, Pa.

Home of two of the greatest bulls of the breed.

Lessnessock King of Beauty 9726 Imported

Lessnessock Douglas Monarch 10020 Imported.

We breed for large production at the pail. Proper type. Beautiful udders and long teats. Note the famous cows in the Herd. Auchenbrain's White Beauty 2d, 21687 Imported, pronounced by prominent breeders the greatest cow of the breed. Advanced Registry test 10,494 lbs. Milk, 511 lbs. Butter. She is increasing this test very much this year.

Garclaugh Bloomer 2d, 20944 Imported, winner of seven 1st prizes in Scotland, and numerous 1st & Champion prizes in Canada. Advanced Registry test 12700 lbs. Milk and 575 lbs. Butter.

Castlemain's Nancy 2d 21686 Imported. Advanced Registry test 11,750 lbs. Milk, 510 lbs. Butter.

Wee Jenny of Holehouse 20952 Imported. Advanced Registry test 10,000 lbs. Milk, 515 lbs. Butter.

Dewdrop of Springhill 20950, Imported. Advanced Registry test, 10,554 lbs. Milk, 457 lbs. Butter.

Primrose 20970. Advanced Registry test, 10,410 lbs. Milk, 450 lbs. Butter.

And many other famous cows.

Bulls to Head Herds, a specialty.

PENNSYLVANIA

Hillview Stock Farm

Paoli, Chester Co., Pa.

"HERD OF REGISTERED AYRSHIRES"

Selected from large producing strains of American and Imported Animals.

Herd headed by Moonstone No. 8419. Sire Moonstone of Drumsuie Imported No. 8228. Adanced Registry No. 7. Dam Flora 3rd. of Bonshaw Imported No. 15575, 22,159 lbs. of Milk in two consecutive years.

Assisted by Dairy King of Avon No. 10733. Sire the Great Show Bull Howie's Dairy King Imported No. 9855, recently sold for the highest price ever paid for an Ayrshire Bull in America.

Dam, celebrated Show Cow Maud Douglas 4th No. 12565.

In this combination of blood I consider I have one of the greatest Breeding Bulls in the country to-day. His get show great udder development, long teats, strong constitution, and great capacity.

Animals of all ages for sale at all times.

Foundation stock a specialty.

W. W. Blake Arkcoll, Prop.

N. B. Paoli 20 miles from Philadelphia on Main Line Penn. R. R. Trains every half hour from Philadelphia. Telephone 101 W. Sugartown.

PENNSYLAANIA

"Ayrmont Farm"

Herd Established 1869　　　　　　　　Tuberculosis Free

The Bulls, Triune No. 11304 and Duke of Netherhall No. 9682, at the head of the herd. Triune's dam, Polly Puss, No. 16296 ex-champion, and sister of Rena Ross No. 14539, now champion Ayrshire cow of the world. Duke of Netherhall's dam Beckey of Holehouse No. 17015 Imported has a 4 year old record of 10,507 lbs. Milk and 463 lbs. Butter. Our herd holds 1st prize for best 5 cows, in Home Dairy Test 1908, and our herd holds 2nd prize, for best 5 cows, in Home Dairy Test 1907.

We have twelve cows, and one bull, in the advanced registry as follows:

Name Cows	No.	Age.	Lbs. Milk.	Lbs. Butter.
Ada Rome	17461	7 yrs.	10,400	424 (retest)
Anna Webb	17454	8 "	9,048	395
Ruth Webb	17457	8 "	8,500	399
Amelia Sebastian	18680	5 "	8,604	414
Pauline "	18687	4 "	10,745	422
Abbie "	20531	4 "	10,449	426
Bernice "	20528	4 "	9,467	415
Agnese "	18679	4 "	9,364	427
Polly "	20527	4 "	8,189	365
Hazel "	20530	4 "	7,825	390
Angeline "	18681	3 "	8,176	397
Chautauqua Fairy	19350	3 "	7,524	356

Bull, Sebastian No. 6269 with eight in Advanced Registry.

STOCK FOR SALE　　　　　　　**PRICES AS TO QUALITY**

Address, J. F. BUTTERFIELD,
South Montrose, Pa.

PENNSYLVANIA AND MARYLAND

Young Stock For Sale

from the Choconut Valley Herd of Ayrshires, got by Gipsy's Pride No. 8955, a son of Cock-a-Bendie Imp., and from cows got by Dandy Webb No. 5462, and Sebastian No. 6269, who have daughters in the Adv. Registry. Come and see them or write to

P. BYRNE & SONS,

St. Josephs, Susq. Co. - - Pa.

Oakwood Stock Farm
AYRSHIRES

Choice young stock for sale.

MacEan Pride No. 10423 from Dr. Butterfield's famous herd, stands at the head of herd.

Returns from Borden's Factory $110.20 per cow fo 1908. Correspondence invited. Prices Right.

ROB'T TEMPLETON & SON, Ulster, Pa.

Borovaccinated and Tuberculin Tested

... AYRSHIRES ...

Herd bull a son of Howie's Dairy King No. 9855
and Miss Hamilton No. 15985.

DR. WM. C. JOHNSON,

Frederick, - - - Md.

WEST VIRGINIA

HILL TOP FARM
AYRSHIRES

Herd headed by Advance Registry sire Nox'emall 7312.

Until recently I used the great show bull and sire Howie's Dairy King (imp.) No. 9855. I am using on my young herd the bulls Madonna's Lad No. 11343, and Kingmaker 11207.

The females in this herd are of splendid type, with excellent udders and teats of good length.

Herd Tuberculin Tested.
All records below given are official.

Nox'emall has sired the following, all but one of which were bred, raised and tested on my farm:—

Name	No.	Age	Milk	Butter
Douglas Daisy Queen	24883	2 yrs.	7,785	322
Douglas Cordelia	21852	2 "	8,874	358
Cordelia Douglas	20326	4 "	8,025*	370*
Douglas Empress	21851	4 "	9,166	441
Babe's Duchess	22213	3 "	9,559	452
Baby Douglas	21843	2 "	9,654†	440†
Daisy Jewess 3rd	20324	3 "	9,665	416
Madonna Lass 3rd	21850	3 "	10,466†	502†
Drummond's Gem	22652	3 "	10,746	416
Etta Poultney	18148	5 "	11,475	489

†Champion records when made.
*Year uncompleted.

The following are also Advance Registry matrons at Hill Top Farm:—

Name	No.	Age	Milk	Butter
Daisy Jewess 2d	17472	4 yrs.	8,574	393
Letta Lind of Radnor	17892	2 "	8,602†	435†
Hillside of Toledo	16022	8 "	9,179	385
Addington Queen	18948	7 "	9,348	456
Madonna Lass 2nd	17473	4 "	10,020	449
October Lass	20323	4 "	10,078	462
Beckey of Holehouse	17015	4 "	10,507	463

†Champion records when made.

Many more cows and heifers will qualify for Advance Registry in 1909.

L. A. REYMANN,

Wheeling, - - West Va.

20180 BELL AYER, Adv. R. 56.
Owned by John R. Valentine, Bryn Mawr, Pa. Record, 11,934 Lbs. of Milk and 575 Lbs. Butter.

20970 PRIMROSE, Adv R.
Owned by Penshurst Farm, Narberth, Pa. Record as a Mature Cow, 10,410 Lbs. of Milk and 450 Lbs. of Butter.

15026 ROSE CLOCKSTON, Adv. R. 27.
Owned by C. M. Winslow, Brandon, Vt. Record as Two-Year Old, 6,135 Lbs. of Milk and 244 Lbs. of Butter.

MARETTA 17488.
Owned by Skyland's Farm, Sterlington, N. Y.

CORRECTOR 9565.
Owned by Skyland's Farm, Sterlington, N. Y.

LADY WONDER'S DAUGHTER 22371.
Owned by Skyland's Farm, Sterlington, N. Y.

FINLAYSTONE 8882.
Owned by Highland Farm, Bryn Mawr, Pa.

20511 ROSE EATON, Adv. R. 80.
Owned by C. M. Winslow, Brandon Vt. Record as a Two-Year Old, 7,783 Lbs. of Milk and 324 Lbs. of Butter.

12094 ACELISTA, Adv. R. 8.
Owned by C. M. Winslow, Brandon, Vt. Record as a Mature Cow, 11,856 Lbs. of Milk and 489 Lbs. of Butter. Five consecutive year record in Home Dairy Test, 52,000 Lbs. Milk and 2,137 Lbs. of Butter and dropped five calves.

SKYLAND'S FARM HERD OF AYRSHIRES
Sterlington, N. Y.

22219 BARLEITH SNOWDROP, Adv. R. 101.
Owned by Barclay Farm, Bryn Mawr, Pa. Record as a Three-Year Old,
8,158 Lbs. of Milk and 349 Lbs. of Butter.

18255 ROSE AILEEN, Adv. R. 51.
Owned by C. M. Winslow, Brandon, Vt. Record as a Two-Year Old, 6,256
Lbs. of Milk and 248 Lbs. of Butter.

21625 AUCHENBRACK SWEET PEA, 2d, Adv. R.
Owned by Barclay Farm, Bryn Mawr, Pa. Record as a Mature Cow,
13,054 Lbs. of Milk and 620 Lbs. of Butter.

17212 FLORINE CORSLET, Adv. R. 40.
Owned by C. M. Winslow, Brandon, Vt. Record as a Two-Year Old,
One year and 335 days old at beginning of test, 5,504 Lbs. of Milk and 248
Lbs. of Butter.

LADY FOX 9669.
Owned by Geo. H. Yeaton, Dover, N. H.

BARCLAY HERD OF AYRSHIRES
Bryn Mawr, Pa.

PAULINE STAR 21090. ADELIA STAR 21089.
MAE OF ELM VALLEY 23779.
Owned by W. G. Tucker, Elm Valley, N. Y.

FIRST PRIZE PRODUCE OF AYRSHIRE COWS.
Wisconsin State Fair, 1907. Owned by Adam Seitz, Waukesha, Wis. This shows the mother and three daughters, all of one type.
Aug. 29, 1908.

BARCLAY FARM.

Queen of Barclay, Brown Eyes of Knockdon, Lizzie of Barclay, Bessie of Rosemont, Friskie of Bonshaw.

20944 GARCLAUGH BLOOMER 2d, Adv. R.
Owned by Penshurst Farm, Narberth, Pa. Record 12,964 Lbs. of Milk and 573 Lbs. of Butter.

CAPTAIN FRANK 9711.
Owned by H. M. Kimball, Concord, N. H.

VIOLA DRUMMOND 12533.
Owned by J. F. Converse, Woodville, N. Y.

HINDA ROSE 18541, DURWOOD'S ROSE 18542.
Owned by Cloverly Farm, Greenfield, N. H.

20950 DEWDROP OF SPRINGHILL, Adv. R.
Owned by Penshurst Farm, Narberth, Pa. Record as a Mature Cow, 10,554 Lbs. of Milk and 457 Lbs. of Butter.

13606 CAD'S BEAUTY, Adv. R. 6.
Owned by E. J. Fletcher, Greenfield, N. H. Record as a Mature Cow, 8,702 Lbs. of Milk and 446 Lbs. of Butter.

21337 GAZELLE MELROSE, Adv. R.
Owned by W. V. Probasco, Cream Ridge, N. J. Record as a Two-Year Old,
9,194 Lbs. of Milk and 407 Lbs. of Butter.

20922 MAY MITCHEL, Adv. R. 156.
Owned by Penshurst Farm, Narberth, Pa. Record as a Mature Cow,
9,419 Lbs. of Milk and 415 Lbs. of Butter.

www.ingramcontent.com/pod-product-compliance
Lightning Source LLC
Chambersburg PA
CBHW062216220526
45471CB00009B/3223